U0024293

回饋訊息於
科技教學的功能與效應

Feedback in
Technology-Based Learning Environments

李宜珍博士（Doris Lee, Ph.D.）著
林浩寬（Andrew Lin）插圖

CONTENTS

CONTENTS

Chapter One
第一章

Introduction
本書詳介

本章要點

　　回饋訊息指的是具有確認、訂正、或是解析效果的資訊。一般說來，回饋訊息都是用以通知學生的學習或考試結果。但它亦可用以解釋分析學生的學習困境。本書詳介回饋訊息的定義、類型、理論依據、及其使用於科技教學與學習環境裡的實証效應。本書並提供相關建議以便設製高效應的回饋訊息。本書共分成五章節。第一章述及回饋訊息的定義及本書的結構。第二章探討並詳介科技教學與學習環境。此章並建議如何設製高功能的科技教學與學習環境。第三章研討回饋訊息源自行為主義、認知主義、及建構主義的眾多研究與結果。第四章細論學習者如何吸收、了解、與傳遞回饋訊息。本章並提供如何有效地於科技教學與學習環境裡提供回饋訊息。第五章總結本書的各項重要論點。

Introduction

Feedback is the corrective or evaluative information which follows and is dependent upon a learner's response. Often times, feedback is applied as an unitary instructional unit and its content can range from the simplest "yes" or "no" message, to the indication of the correct answer, and eventually, to substantial corrective information or remedial instruction (Carter, 1984; Clariana & Lee, 2001; Kulhavy, 1977; Lee, 1994; Lee & Dwyer, 1994; Lee, Savenye, & Smith, 1991; Mory, 1991, 1994, 1996). Feedback has been perceived as critical to increase the effectiveness of instruction, however, "it is one that is too often slighted or overlooked" (Smith & Ragan, 1999, p. 119). In recent years, feedback has been widely integrated into all kinds of technology-based instruction and learning including computer, internet, intranet, and web-based modules. Again, unfortunately, not much well-organized and systematically synthesized information on feedback is available.

The purpose of this book is to provide a comprehensive review of feedback, its functions, mechanisms and effects in technology-based learning environments. Technology-based learning environments refer to the many different definitions and categorizations existing for the various forms of using a computer, a public computerized network such as the Internet, a private computer network, or a company's intranet, to deliver a learning module or lesson (Curtain, 1997; Lee & Borland, 2007; Lee, Frenzelas, & Anders, in press; Chalmers & Lee, 2004; Lee, Chalmers, & Ely, 2005; Trombley & Lee, 2002). In addition, this book aims at offering insights and suggestions for the effective integration and application of feedback into these modules.

This book contains five chapters. Chapter one introduces the intent and basic organization of the book. Chapter two includes an introduction of the many different definitions and categorizations for technology-based learning and related design issues. A technology-based learning environment includes e-learning, web-based learning (WBL), computer-based learning (CBL) or computer-supported learning (CSL) (Lee & Borland, 2007; Lee, Frenzelas, & Anders, in press; Chalmers & Lee, 2004; Lee, Chalmers, & Ely, 2005; Trombley & Lee, 2002). Chapter three focuses on feedback's definitions, types, functions and effects. In this chapter, the role, influence and effects of feedback in a technology-based learning environment will be emphasized. Chapter three also reviews pertinent studies that investigate feedback effects or use feedback as a critical variable to improve learning in print-based, instructor-led and technology-based learning environments. Insights and directions for the effective application or use of feedback in technology-based environments will be discussed in this chapter as well. Chapter four provides discussions as well as examples regarding the processing of different types of feedback information. This chapter further discusses future trends concerning the integration of feedback in technology-based learning modules and provides suggestions. Finally, Chapter five concludes this book.

In summary, systematic, comprehensive studies of how feedback works in the emerging technology-based learning environment are very limited. This book intends to contribute to the feedback literature by unveiling the mechanism, and the effects of feedback. Finally, this book is significant in that it will reveal, synthesize and discuss empirical results on how feedback functions to improve learning in various types of technology-based learning environments including a stand-alone computer-based or computer assisted lesson and the emerging wireless, virtual classrooms where learners can simultaneously work on the same file and are able to edit, update, or post information in real time.

Chapter Two
第二章

Technology-Based Learning Environments:
Definitions, Important Issues and Suggestions
高效應科技教學與學習環境的定義、重要課題、及建議

本章要點

　　正如前言，此書的目的在於探討回饋訊息使用於科技教學與學習環境的的定義、類型、作用、機制和效應。本章探討何為高效應科技教學與學習環境並論述具有相關性的重要課題及建議。本章涉及眾多議題，這些議題包括：電腦軟、硬體、多媒體、掌上型電腦及網際網路。本章並論及面對面的學習環境、電子教學環境和綜合面對面與電子教學環境的學習情境。除此，本章並述及設計這些課程的考慮要素及設計重點。最後本章以建議如何設計高效應的科技教學與學習環境為結語。

Introduction

The focus of this book is on feedback, its definitions, types, functions, mechanisms and effects. This chapter serves to define what a technology-based learning environment is, the many factors that need to be considered in order to design an effective one and pertinent suggestions. Based on this premise, this chapter tries to cover pertinent topics of using computers and their related technologies in a learning situation. The first section covers the frequently used technologies including computer hardware, software, the internet, the web and emerging content management systems. Following that, three different types of learning environments including face-to-face, e-learning and blended types of learning involving the other two types of learning will be explained. The role that technology plays in each of these three types of learning environments will be discussed. Then, this chapter will continue to examine the many issues facing technology design, integration and implementation. Suggestions regarding how to effectively integrate technologies for the best learning results and how to systematically design and implement technology-based learning lessons will be discussed. Finally, a conclusion that synthesizes the key points of this chapter will be provided.

Defining a Technology-Based Learning Environment

The computer revolution has resulted in the rapid development and proliferation of technology in homes, schools and companies in the past few decades. The people of most developed countries are now using numerous kinds of cutting-edge technologies to conduct daily tasks or to engage themselves in information gathering and delivery. Accordingly, how to create an effective technology-based learning environment has become one of the critical issues in many educational and business organizations. The focus of the book is mainly on feedback functions and effects. The additional focus of this book is on the process that learners take to process feedback information and how to effectively integrate feedback into a technology-based learning environment. It is then necessary to discuss what constitutes an effective, technology-based learning environment.

In the past, radio, television, films, calculators, slides and overhead projectors were commonly referred to as "media" or "technology". In the late 80's and early 90's, with the rise of the personal computer in the context of learning, the term "technology" began to associate with computers or more precisely, personal computers. Terms or categorizations like computer-assisted instruction (CAI), computer-based instruction (CBI) and computer-managed instruction (CMI) were widely employed to describe how a lesson was presented to a learner through a stand-alone personal computer. These terms are still employed by researchers or practitioners to loosely describe the use of personal computers to deliver learning or instructional lessons. However, during all these years, it has been difficult to find standardized definitions, categorizations or classifications for the various forms of technology-based learning. Some of the previously used terms may no longer be appropriate to describe the current status of using the internet or the World Wide Web (WWW) or simply called the web, for instructional or learning purposes. In recent years, personal computers are rarely "stand-alone"; most often, personal computers are now linked to other computers, computer networks or additional technologies. Due to this reason, in this book, the terms like computer-assisted, computer-based or computer managed learning are not used; rather, the term "technology-based learning" is employed to cover the latest computer and computer related technologies.

Again, currently, reality has dictated that a very long list of computers with additional, related technologies and peripherals have been commonly used as learning media in both school and business settings. Still, no matter how long and how extensive this list is, fundamentally, in a technology-based learning environment, technology should never be the primary catalyst for learning. Technology should be there for one purpose only: to maximize the learning or instructional results. In this book, a technology-based learning environment is defined as a learning situation or environment where many different forms, types, or modules of computer related systems and/or technologies are used in order to satisfy, facilitate or enhance a human and/or learning need. For this purpose, in a technology-based learning environment, all the needed knowledge, tools, theories, paradigms, and/or skills could be employed or involved simultaneously with technology used (Bober, 2001).

In other words, a technology-based learning environment does not merely include a certain type or form of technology; instead, it could involve or employ various forms of technology, various types of media and many different schools of paradigmatic theories and knowledge bases to fulfill a learning need. Also, how effective a technology-based learning environment actually is should not simply be evaluated by the amount or the type of technology used. Rather, it should be judged by how well the lesson provided in the environment can utilize all involved to achieve the best learning results (Chalmers, Lee, 2004; Choitz & Lee, 2006; Lee & Borland, 2007; Lee, 2004; Lee, Chalmers, & Ely, 2005; Trombley & Lee, 2002).

Graphic 1: An Example of a Technology-Based Learning Environment

Frequently Used Technologies in Technology-Based Learning Environments

Again, the list of available technologies that could possibly be used in a technology-based learning environment could be overwhelmingly lengthy. However, terms or categories including computer hardware, software, multimedia, internet-based learning, web-based learning, e-learning, m(mobile)-learning, content management and learning content management systems are the ones frequently used to describe various kinds of computer-related technology employed for learning or instructional purposes. Below covers each of these terms.

9

Computer Hardware, Multimedia, and Handheld Computers

Most of the terms covered here are, by now, common knowledge. Thus, the purpose of this section is to provide a brief, concise overview regarding possible hardware options in a technology-based learning environment. The information of this section is a synthesis of numerous commercial publications on modern computer technology and books published by Clark and Mayer (2003) on e-learning, Khan (1997) on web-based instruction and Vaughan (1996) on multimedia. Briefly, computer hardware consists of the physical components of a computer. In the eyes of most, computer hardware is referenced to as the hardware in a computer system, not as in the computers that govern such machines as ATMs, car engines, electrocardiograph machines, CD players, and other devices. Computer hardware can be separated into three distinct categories: input devices, output devices, and storage devices. Input devices are parts that allow a user or a learner to input information or commands into the computer's memory. Some examples of input devices are the mouse, keyboard, a microphone, a digital camera dock, and a PC camera. A keyboard allows a learner to input letters, words, numbers, and symbols into the computer's hard drive. A microphone lets a learner say commands that the computer will follow or lets the learner enter data verbally. A mouse is a small handheld device that is used to control a symbol known as a pointer. The pointer moves across the screen, and selects programs to run. A digital camera dock allows a learner to store images in the computer's hard drive. PC cameras allow a learner to place videos into computer storage.

Output devices are devices that synthesize information from the computer into a format users can understand. Commonly used output devices are a printer, monitor, and speakers. Printers produce information on a hard-copy medium. The monitor is like a television screen which displays information, and speakers allow learners to hear sounds, words, and music from the computer. The system unit has a case or chassis in a tower shape with additional parts, including the following:

- motherboard or system board with the central processing unit (CPU) where most of the work of a computer system takes place,
- random access memory (RAM) for program execution and short term data storage,
- read only memory (ROM) that contains information for a computer system to function properly,
- basic input-output system (BIOS) or extensible firmware interface (EFI) in newer computers,
- buses for transferring data or power between computer components inside a computer or between computers,
- power supply, and the storage controllers that control the hard drive, floppy drive, CD-ROM (Compact Disk-Read Only Memory) drive, USB (Universal Serial Bus) drive, video display controller,
- Computer bus controllers to connect the computer to external peripheral devices such as printers or scanners.

In addition, some type of a removable media writer include commonly used compact disk (CD), the most common type of removable media, CD-ROM Drive, CD writer, DVD (Digital Video Disc) , DVD-ROM (Digital Video Disc-Read Only Memory) drive, DVD writer, DVD-RAM (Digital Video Disc-Random Access Memory) drive, floppy disk, zip drive, USB Flash Drive, tape drive for backup and long-term storage. Internal storage that is used to keep data inside the computer for later use include, CD-ROM Drives, CD writer, DVD, DVD-ROM drive, DVD writer, DVD-RAM drive, floppy disk, zip drive, USB Flash Drive, tape drive for backup and long-term storage, hard disk, disk array controller, sound card, networking that connects a computer to the Internet and/or other computers, modem for dial-up connections, network card for DSL/Cable internet, and/or connecting to other computers.

Hardware can include external components of a computer system including:

- wheel mouse,
- input or input devices,
- text input devices, keyboard,
- mouse,
- trackball,
- gaming devices, joystick, gamepad, game controller,
- image, video input devices, image scanner,
- webcam,
- audio input devices, microphone,
- output or output devices image, video output devices,
- printer peripheral (Inkjet, Laser) device for printing a hard copy.

Additionally, computer hardware also includes monitor devices that take signals and display them such as CRT (Cathode Ray Tube), LCD (Liquid Crystal Display), audio output devices like speakers and headset that are similar to a regular telephone handset but are instead worn on the head to enable hands-free operation. Finally, the term, multimedia, is often used to represent the multiple forms of information content and information processing (e.g. text, audio, graphics, animation, video, interactivity) performed by a computer system by using some or all of the hardware components introduced.

Graphic 2 : Technology

Recently, handheld computers are gaining popularity. Handheld computers, which are informally referred to as "handhelds", are personal digital assistants. These assistants can be placed or held in a person's hand and can be operated from a main power source. Handhelds were originally known as Palm Pilots, as they were from the Palm Company and have different labels including digital assistants (PDAs), palm computers, palm pilots, mini-notebooks, or pocket personal computers (PCs), pocketPCs. Handhelds are not as expensive as desktop or laptop computers, but they can be used as a replacement for the personal organizer with features including a calendar, an address book and other simple data storage applications. It is expected that handhelds will be merging with mobile phone technology either as phones with PDA facilities or PDAs with phone applications (Perry, 2003; Clyde, 2004; Polsson, 2005; Lee & Riordan, 2007).

Graphic 3 : Handheld Computers

Recently, new technologies have emerged enabling the common chalkboard to be supplanted by interactive whiteboards. Interactive whiteboards use special connection software to enable instructors to connect their computers with their interactive whiteboards, enabling the presentation of digital media stored on the instructor's computer, such as PowerPoints, software, Spread-sheets, graphics, and video clips. Furthermore, many interactive whiteboards feature a write-on feature, in that they possess interactive pens and erasers, turning the Smartboard into a combination of a chalk-board and digital projector. The digital whiteboard can also transform hand-written text into type. In addition, the Smartboard can be compared to a digital notebook.

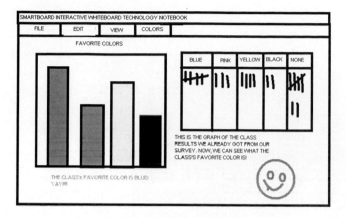

Graphic 4: A Smartboard

Computer Software or Courseware

As opposed to the physical elements of a computer, computer software refers to series of changeable, modifiable and erasable instruction or computer programs that direct the computer hardware to execute jobs. As Lockard, Abrams, and Many (1990) explained, software is the element in a computer system that makes the entire system work. Software directs the CPU with individual instruction as to what should be performed; the control section of a computer system then directs and coordinates all the necessary parts to execute the instructions. A computer program, which is more commonly used by computer science professionals and often interchangeably named as computer software, also contains a series of instruction for a computer system to produce the intended results. Programming or coding refers to the writing of these programs, and computer programs or software could be purchased, developed in house by users or companies.

Specifically, in the context of instruction or learning, since the early 90's, the term computers as tool, tutor or tutee, coined by Lockard, Abrams, and Many (1990), has been widely employed to describe what and how software can be used for instructional or learning purposes. According to these researchers, primarily, software that can be used a tool include application software, hypermedia and integrated type of software. Application types of software include word processors, data management and filing programs. Hypermedia or hypertext are authoring type of programs that enables a user to create programs with text, graphics, audio, video, animation, film clips and links between screens. The integrated software include the programs that integrate multiple functions such as word processing, data management, telecommunications and desktop publishing for the production of graphics, classroom publishing, emails, and other supporting materials.

Using instructional software as a tutor occurs in learning situations where the instructor arranges computers and necessary software for learners to engage in meaningful and needed activities and interactions between the software and the learners. Commonly used programs in this category include tutorial, drill and practice, simulations and instructional games (Alessi & Trollip, 1991; Lockard, Abrams, & Many, 1990). Specifically, tutorials are used to provide learners to learn new and unfamiliar content, and there are linear and branching types of tutorials. In a way, tutorials are core lessons that give the learner new or unfamiliar information. Drill and practice helps learners to practice on previously learned content through the "drills" provided. These drills emphasize, in many cases, speed. Simulations include simulated yet realistic situations for learners to manipulate related variables. Learners are often presented with realistic scenarios, are required to analyze the variables, factors or situation involved in the situation in order to make

Graphic 5 : Software

decisions or to solve related problems. Instructional games present learning content in a game-based format. Instructional games could be perceived as simulations with the competition element.

Internet-Based, Web-Based Learning, E-Learning or M-Learning

The definitions and classifications of software and its use date to the 1990's. However, the focus of technology in the 2000s, rather than being the use of software, is the use of networks ranging from the global internet to small company intranets. In the 90's, the internet, a gigantic information delivery, exchange and transfer system, began to gain popularity (Engle, 1999). In contrast to the massive global internets that are used by billions of people every day, companies, organizations, education workers, and associations use intranets, private networks where group technology is used.

In common usage, people tend to use the terms Internet and World Wide Web as if they were the same. On the contrary, as Shelly, Cashman, Gunter, and Gunter (2004) explained, the World Wide Web and the Internet are not the same, but are instead similar to a part to a whole. The World Wide Web is a subsidiary of the actual Internet. The internet predates the World Wide Web by 30 years, having started as a military tool. The World Wide Web, designed at the European Particle Physics Laboratory in Geneva, Switzerland, by Tim Berners-Lee, is a subset of the Internet, designed in the early 1990's. The World Wide Web gained in popularity rapidly, and quickly overtook the internet in the number of users. The fact that information can be transferred quite quickly to and from users has become the primary factor in the World Wide Web's popularity.

The modern terminology designed to supersede the older software-based terms used in the 90's stem from the extensive use of the World Wide Web. Most of the terms that are now in common use have evolved from the terms formerly used in the 90's. Standardized categorizations for internet learning are even harder to impose than software categorizations, as internet learning is constantly advancing at the waves of millions of people developing and using programs based on it every day.   The following list is to clarify some of these terms:

- E-Learning or M-Learning: E-learning or m-learning is the practice of learning via technology, which can include anything from using a one-on-one chat room to a massive internet forum, with the format of using technology in a classroom or in a mobile format.

- WBL: WBL, otherwise known as Web-Based Learning, is learning via the internet. Technologies used in WBL can include anything from text-based web pages and files, graphics, and videos to simulations, interactive web conferencing, chat rooms, and threaded discussions.
- Synchronous WBL: A subdivision of WBL, synchronous web-based learning involves situations in which learners and instructors are interacting simultaneously, moving at a set pace determined by the instructors. Examples of this include online classrooms, internet forums, and chat rooms.
- Asynchronous WBL: Like synchronous WBL, asynchronous WBL is a subdivision of WBL. Asynchronous WBL allows learners to proceed at a pace that moves as quickly or as slowly as appropriate. Examples of this include discussion groups, email conferencing and video conferencing, which allow learning without the interference of other learners (Clark & Mayer, 2003; Khan,1997).

It needs to note that even with the fact that conversations may be extended over days, weeks, and even months, both synchronous and asynchronous WBL allow for conversation between all group members, thereby establishing that both can reasonably hold learners from all parts of the world. Simulations realistically depicting actual situations can be used and have great potential in helping facilitate connection development. These can also greatly increase peer-to-peer interaction and help increase knowledge stores (Bonk & Reynolds, 1997; Chalmers & Lee, 2004; Choitz & Lee, 2006; Harasim, Hiltz, Teles, & Turoff, 1995; Lee,

Chalmers, & Ely, 2005; McCreary & VanDuran, 1987; Palloff & Pratt, 1999).

A Learning Management System (LMS) and a Learning Content Management System (LCMS)

In recent years, LMS and LCMS systems have been used by companies to be able to examine training practices. LMS systems are used to primarily keep records of learners and view their general status over periods of time. These are often used by corporations to monitor trainee employee status, and are often also used to help in troubleshooting training systems. LMS systems can display anything from:

- Time
- Place
- Format
- Learning System Used
- Classroom Deliveries
- Downloadable Documents

LMCS systems, in contrast to LMS systems, manage content rather than learners, and are commonly used to help with the creation, management, delivery, and other factors regarding curricular activities. SCOs, known as Sharable Content Objects, contain learning materials, and can be anything from documents to videos to web pages. The combined use of LMS and LCMS can help greatly in managing learners, developing new content, maintaining comprehensive control and a trouble-shooting system for company training programs (Lee & Borland, 2007).

Technology Usage in Different Types of Learning Environments

According to Lee, Frenzelas and Anders (in press), three major types, formats or classifications of learning environments exist: Instructor-led (face-to-face), e-learning and a blended learning one in which the two preceding types are combined together, all exist. The following explanation is based on Lee et al. with reference to each of the three types or formats of instruction in-depth.

An Instructor-Led, Face-to-Face Learning Environment

Commonly, instructor face-to-face environments occur in classrooms in schools or in conferences for training. Micro and macro strategies must be used by instructors if they wish to successfully educate and satisfy learners. Instructors use macro strategies by striving to inform learners of the objectives, the overall "big picture", and to gain increased motivation. Micro strategies focus more so on group discussions, case studies, more so "specific work". Thus, lectures, overhead projectors, computer-generated PowerPoint presentations, and interactive whiteboards are the instructional methods and technologies often used to gain more information regarding the topic. This type of instruction is characterized, most often, by a general set pace, where one moves at a steady rate (Dick, Carey, & Carey, 2005). The events of instruction often include the nine ones proposed by Gagne (1970, 1974, 1977) for the best results. Again, at any point of time for any event, one or a combination of the aforementioned technologies could be used to enhance the instruction effects. These events are:

Graphic 6: Self-Paced Technology-Based Learning

- calling learners' attention to the to-be-learned content by using interesting stories, graphics or jokes;
- informing learners the learning objectives;
- help learners recall the relevant, previously learned content;
- present relevant, stimulating learning content and materials;
- offer needed learning guidance to learners;
- encourage learners to practice and perform;
- assess learner's performance;
- provide feedback concerning the success of the performance;
- help learners enhance and transfer of the learned content.

An E-Learning or M-Learning Environment

E-Learning, or m-learning (e-learning in a mobile format), the practice of learning utilizing the internet or another system of networked computers, can be classified as CBL and WBL learning. With e-learning or m-learning, technology is paramount for operation, with video and audio downloads, simulations, interactive games, bulletin boards, text-based files, web pages, graphics, and numerous other tools all featuring prominently in the technology necessary for the operation of e or m-based learning curricula. E-learning or m-learning also can include some or all of the nine events listed above as a design focus, enabling designers to take these factors into account. As stated previously, synchronous e-learning or m-learning occurs when technologies enable real-time interactions, such as chat rooms, forums, and video conferencing. On the contrary, asynchronous e-learning or m-learning is more so individual based, with the instruction tailored to the learner's pace by using threaded discussion groups, emails, and bulletin boards (Chalmers & Lee, 2004; Clark & Mayer, 2003; Khan, 1997; Lee, Chalmers, & Ely, 2005; Lee, Frenzelas, & Anders, in press; Lee & Borland, 2007; Macpherson, Homan, & Wilkinson, 2005; Nisar, 2004; Trombley & Lee, 2002).

A Blended Learning Environment

Theoretically, blended learning combines the best qualities of face-to-face and e-learning and melds it into one great pot of advantages, with everything from the personalized aspect of a face-to-face classroom situation to the efficiency of online systems. In other words, blended learning, unlike the former two types of learning, therefore, may include different alternative methods, providing much variety for the learner and the instructor to choose from, including anything from online self-paced training to instructor-led classroom learning combined with e-learning. Like e-learning and face-to-face instruction, some or all of the nine events mentioned above can be used. There are numerous examples that could be used. One is a learning program in which learners first use a self-taught style learning program, possibly coupled with an online instructor to help with technology problems or difficulties. The second part of class would have learners review the content by possibly performing a project or a classroom event, then receiving feedback for their efforts. Drilling would be performed, and then post assessments such as tests or graded homework assignments would be assigned. After the class ends, learners can discuss and trade information regarding the classroom session, and can have in-depth review time for the next session. In this way, effective learning can continue, even with the influx of new technologies now available to instructors (Lee, Frenzelas, & Anders, in press).

Graphic 7: Blended Learning

Critical Factors to Consider for Using Technology-Based Learning Lessons

Researchers have continued to point out that when an organization, school, group, or association decides to employ or purchase technology-based learning systems for training or learning purposes, numerous factors must be taken into account. These factors can be further organized into three categories include: financial expenses and returns, content and technical issues, and learner factor. These factors are a result of synthesizing most of the work published by Chalmers and Lee (2004), Clark and Mayer (2003), Driscoll (1998, 2000), Hall (1997), Khan (1997), Lee, Chalmers, and Ely, (2005), Lee, Frenzelas, and Anders (in press), Lee and Borland (2007), and Trombley and Lee (2002). It needs to be noted that this synthesis is an attempt to provide an organized way for looking into the related factors concerning the use of a technology-based learning environment.

Expenses and Returns

The required expenses or cost for a technology-based course or lesson is one of the main factors in its success. The actual expenses, hereby known as the cost, must be reciprocated through an efficient, and high-quality, long-lasting learning experience. One of the primary factors to consider in the design of technology-based courses is whether the module should be built in-house or outsourced to a vendor service. Both have their benefits and disadvantages specific to each other. If the learning content is such that an outside source can be used, outside sources are generally less expensive to use than designing the modules in-house. In-house building, although tailor-made to the exact specifications necessary, can take years to build and operate, with extensive records detailing everything from the interfaces to the animated graphics required. With an in-house design, the costs can range from 75,000 to 1.6 million dollars. Typically, fees for access, licenses, and numerous other factors are required in order to help alleviate some of the cost. Flash, Shockwave, Adobe Acrobat, and numerous other online plug-ins required for the operation of online-based resources such as audio and video clips. Therefore, cost is extremely important. For instance, if the class is being used by a small group of learners who will only actually use it for a very short period of time, the cost of possibly over 1 million, not to mention the time and effort, would have not been spent wisely.

Content Design and Technical Issues

Content and technical issues are important considerations, along with cost, to be remembered and included in plans for WBT servers. With these issues, the challenge is to determine whether or not the content to be presented is to be delivered into a format suitable for all the technologies used via a technology-based format. Question that must be answered before designing the content include: Does the content need to stay consistent over time? Often, the content delivered via a technology-based lesson is basic content which can stay consistent for long periods of time. Updates are not needed unless one of these two events occurs: (a) massive company restructuring, causing change in training or learning systems; and (b) new technologies render old ones obsolete, causing changes to basic training systems. Thus, when these events occur, the technology needed must remain available. Once updates are completed, they can be sent instantly via internet to all parts of the world with internet connections. However, content with numerous facets which require much detail and elaboration would be difficult to use or be constantly updated on a technology-based lesson.

In addition, some of the problems that can be encountered when dealing content delivery include bandwidth limitation and multimedia intensive features. If bandwidth is lower than required for the technology-based lesson, then the lesson will act slowly, delay responses, and cause general trouble. The server, the receiving computer, and the transfer process can all have problems. Multimedia with audio-video intensive features can often slow down bandwidth, causing intermittent stops and starts. These problems can sometimes be alleviated using applets, however, effective and timely technical support is often necessary and, in some cases, required.

Learner Factor

In addition to cost, technical support, and consistency of content, it is also necessary to consider many factors concerning learners. For example, learners need to have a higher level of technological literacy in order to learn well. In addition, because of the lack of face-to-face interaction, it is often difficult for online instructors to develop the rapport with the learner necessary for true control. Finally, due to the impersonal aspect of human-machine-human learning, interpersonal relations between the instructor and the learners, widely considered as an essential component of learning, could be lacking.

Suggestions for Designing Effective Technology-Based Learning Lessons

With the aforementioned considerations, many suggestions are offered, and these suggestions are all centered on the design process, the learners, and the content delivery process. Each of these suggestions are elaborated on in detail below (Chalmers & Lee, 2004; Lee, Chalmers, & Ely, 2005; Lee, Frenzelas, & Anders, in press; Lee & Borland, 2007; Trombley & Lee, 2002).

Suggestions for a Successful Design Process

Numerous challenges exist during the design process. Finding a design model is the first one: traditional design models are mostly for a face-to-face learning style. There are numerous models created for a computerized or technology-based system, however, a standard model is not yet available. Even so, the design process should include phases such as a needs analysis, goals and prioritization, task and learner analysis, objective writing, materials design, evaluation and revision. These phases can all come from a step-by-step process or a rapid prototype acceleration process. Most importantly, a systematic design approach requires constant, on-going evaluations, with the capacity for revision and revamping if necessary, until the module is effective (Dick, Carey, & Carey, 2005).

The needs analysis is the first and crucial step in a systematic approach. Needs analyses check what the technology-based lesson needs to perform and what standards of quality, what types of training or learning, and the kinds of feedback needed? A needs analysis determines the company or organization's specific needs and combines them into one coherent standard that the resulting module or lesson must aspire to. Needs analyses also check for the overall level of literacy with the content to be covered.

Next, design teams with strong technology background are necessary. Strong teams are necessary to put together the jigsaw of a design process. Teamwork is imperative, and strong design teams need numerous types of talents. However, one of the most important is the facilitator. The facilitator can be a designer, site editor, or web master. With the facilitator, the questions of learners can be answered by a person who knows and understands the full lesson. This in turn can make it such that there is communication between learners and designers, allowing for learners to act as critics and suggest improvements. Learners can also help troubleshoot for glitches and problems that might exist.

Graphic 8: A Strong Team

Understanding Learners

Learner analysis is one of the most potent tools for troubleshooting of a technology-based lesson, and can greatly help in evaluation. Understanding of who the learners are, their needs, and their general technological literacy is essential. For example, learners' technological literacy is an important criterion, for if learners cannot operate the lesson, then the learners will most likely, out of frustration, turn off the machine. Learners' data can be collected by a variety of methods such as questionnaires, opinion polls, and/or tests. The information to be gathered includes technical difficulties, academic motivation, attitude towards the learning content, overall perception of the program. For instance, the primary motivation for learners when using a technology-based course is to play a successful role in the learning process (Gottschalk, 1999; Kilby, 1997; Ward, 1998). It is therefore evident that learners must be kept in a high motivation level. Therefore, the lesson must be designed in ways that take advantage of the learner analysis data collected to aid in increasing learners' motivation in the content provided.

Furthermore, another challenge is how to keep learners interested in the content for the duration of the lesson. Numerous researchers have stated that motivation can come from material things, such as a salary increase, job security, and pension, but also can come from an internal determination to learn and continue the endless journey of accumulation of experience and knowledge. Also, such factors as success, volition, value and enjoyment all have their places in motivation (Wlodowski, 1985). These factors, according to Lee, Chalmers and Ely (2005) can be motivating in technology-based learning environments for both young and old. An example would be that if the lesson presents content in a matter such that is relevant to the learners' previous background knowledge, the learner will be better motivated to the lesson. It is also suggested by Choitz and Lee (2006) that the instructor must be fully invested in the learning process. However, Choitz and Lee also stressed that the instructor must not interfere in the learning process to a point at which the learning process is completely under the control of the instructor. The instructor's role is to set expectations, ask questions, provide feedback and keep control over the learners.

Suggestions for Effective Content Design and Delivery

Content analysis, otherwise known as an analysis of the nature and difficulty level of content being taught in a lesson, determines how to chunk the content into what might be termed as "bite-size" learning pieces. Chunking the content and decisions regarding this must be made based on the kind of content being taught. For instance, graphical demonstrations of motor skills can be covered in video. For mathematics skills, an interactive "online chalkboard" video showing the full process might be in order. Drilling and practice must be provided as a primary supplement. Most importantly, well-chunked content is well analyzed and placed into tiny modules which can be used at whatever speed merited. It needs to note that with technology, flashy graphics, animations, and videos are all possible, able to increase prospective interest in the lesson. With a technology-based lesson, learners are no longer restricted by a human instructor. Therefore, special attention must be directed towards keeping learners focused on the content and not on separate possibilities.

Furthermore, learners' demands and their own constraints give a great challenge to designers using technology as a medium of learning.

With numerous factors ranging from boredom to technical problems, constant temptations exist for learners to draw out of the module. Certain strategies that have suggested for keeping learners inside the module with the lesson include:

- Giving increased interactivity by permitting learners to control the order of video and audio clips.
- Giving learners a sense of belonging with interpersonal activities such as debates and discussions.
- Include well-designed assignments.
- Provide detailed, explanatory or informative type of feedback which serves as a messenger stating levels of success in response to the activities performed. Extensive feedback directly from the instructor when dealing with difficult concepts should be given. This type of feedback can serve as a remedial or review lesson for the learner to self-correct his/her misconceptions or errors.
- Offer a variety of activities including games, role play, simulations, debates, polls, and projects. These are very helpful in engaging learners in the content.

In summation, a designer of a technology-based lesson must thoroughly analyze and chunk content into manageable parts which can be quickly and easily swallowed by learners and processed into appropriate knowledge. Learners should have increased options for controlling their learning experience, from being able to stop and start their video segments, and be able to review and repeat information. Also, forums should be given so that learners can comment, report technical problems, and give suggestions for improving quality.

Furthermore, the interface component of a technology-based course determines the outlook learners have on lessons delivered via technology. Arrows, links, and hyper-links should be given easy access, and important components should be presented quickly and simply. The interface should be simple so that major components such as settings, start, and help can be accessed quickly and easily. As a precaution, all pages should have a redirect button to the main page for easy navigation. Many designers have suggested a site map of sorts to help guide learners with all necessary information included, such as course length, course content, and a manual of sorts describing how to proceed through the activities provided for the user's purposes.

It is also advocated that related information such as assignments, grading policies, and instructor expectations should be provided. Furthermore, this site overview should be accessible at all times, with a hierarchical organizational system based on page importance. Help and tech support pages would be easily accessible, and would greatly aid students. With feedback and dialogue boxes which learners could use to submit comments and concerns to the webmasters, satisfaction can be granted. Once again, a forum is essential, allowing learners to post questions and comments. Extensive, informative type of feedback also can give site maintenance a great help. Also, standard browser functions should be kept as records so that companies can keep employees on the same versions of hardware and software, with no compatibility problems. Therefore, it is necessary that compatibility between older and newer versions is established, making it easier for companies to keep track of and manage a learning systems. Other features that would be considered as desirable include glossaries, site maps, and numerous other resources (Clark & Mayer, 2003; Jones & Okey, 1995).

In addition, in order to effectively and truly manage the access and development of the learning content, LMS and LCMS systems could helpful. For instance, LMS data can be found through interfaces which include data from sources all over the company. LCMS systems can help manage content and contain features which can be used to create, reuse, locate, deliver and manage the content. LCMS can also be used to track each learner's use of content. Typically, the cost of a combined LMS/LCMS is stipulated either by a per user fee or a site license fee. LMS/LCMS systems require two years for installation, and often, experienced vendors in LMS/LCMS business are preferred. Vendors should provide LMS loaded with features, and an LMS with the ability to be integrated with other department's data (Lee & Borland, 2007).

Finally, after the plan and design process have gone through, the prototype must be constructed. This module should be tested by numerous people most likely to be similar to actual testing subjects. Testers should be encouraged to point out all flaws in the program they see, ranging from faulty links to bandwidth problems. As soon as the testers clear the prototype for use, the format should be applied to all other modules in the same course (Beer, 2000; Kruse & Keil, 2000).

Lastly, evaluations must be conducted throughout the use of a technology-based lesson. These evaluations allow for updating and troubleshooting of the lesson. As soon as the prototype is complete, formative evaluations should take place, and stimulative evaluations should commence after the course has run its way through. Immediately, all the evaluation data should be used for revision of the module (Beer, 2000; Dick, Carey, & Carey, 2005; Driscoll, 1998; Kruse & Keil, 2000; Hall, 1997; Hanke, 1997; Kilby, 1997; Ward, 1998).

Summary

The development of technology-based lessons is increasingly important in today's economic structure. With the prevalence of technology all over the world, numerous schools and multi-national companies are using all kinds of technologies to conduct training and student learning. This chapter begins with a detailed introduction of the potential hardware and software for a technology-based lesson. This chapter then covers many different factors that must be taken into consideration. Cost-effectiveness, technology, bugs, employee technology literacy, and resources allotted could outweigh the benefits of a technology-based course. The participating learners need to have technical literacy that is at least proficient in order to effectively use the lesson. Technology resentment by learners can cause bad learning spirit, and therefore the participating learners need to be prepared with a proper level of technology literacy and learning motivation. In addition, technology requires constant updating and maintenance, and can sometimes fail. Enormous amounts of internal resources are required. All members in the design and support teams and other staff are all needed to keep an effective technology lesson going. Technical experts, designers, and outside vendors are often needed too. A careful balance between face-to-face instruction and an e-type of lesson needs to constantly be maintained. Concerns about the lack of inter-personal training when utilizing technology for learning or instructional purposes have not been offset, and may be justified. However, many companies show great increases in profits with technology. Therefore, the issue for a company to consider would be how to use technology to help save millions on travel costs for trainers. Furthermore, with a technology-based lesson, the questions regarding the use of facilitators to monitor learners have become issues which need resolution.

Based on these factors and concerns, this chapter provides pertinent suggestions to raise the level of effectiveness of a technology-based lesson. These suggestions include employing a systematic design process, involving strong design teams, attending to learners' needs while learning with technology, integrating thorough content analysis and design, including appropriate interface design, considering using learning management systems and/or learning content management systems, and incorporating piloting and evaluation to ensure the quality and success of the lesson. However, certain questions are beyond the scope of the book; they remain to be answered by future researchers. For example, what, content or technology available, should drive the design of the lesson? What would be the best blended lesson for whom? Finally, in a technology-based lesson, what type of instructional or learning variables are absolutely necessary? One of the variables is feedback. Feedback that can be designed to inform the learners about their learning progress, mistakes or achievement is crucial for any type of technology-based lessons. The next two chapters will focus on feedback: definitions, types, different theoretical background, effects, processing procedures and suggestions for the effective integration of feedback.

Chapter Three
第三章

**Definitions, Types, Theoretical Bases,
Functions and Effects of Feedback
in Technology-Based Learning Environments**
回饋訊息的定義、類型、理論依據、
及其使用於科技教學與學習環境裡的實証結果及效應

本章要點

　　本章探討回饋訊息的多項要件及其使用於科技教學與學習環境裡的實証結果。正如前言，本書的目的在於研討回饋訊息於科技教學與科技學習環境裡的功能與效應。前二個章節的重點在於科技教學與學習環境或情境的介定。前二個章節並探討如何設計有效的科技教學與學習環境。本章更進一步探討有關回饋訊息的諸多細節。這些細節包括回饋訊息的定義、類型、理論依據、及其使用於科技教學與學習環境裡的效應。諸多過去與較新近的實驗及研究都將於此章節細論。眾多涉及並源自行為主義、認知主義、及建構主義的研究、理論及其實証結果與發現都將於此章一一論述。

Introduction

The main purpose of this book is to discuss both the functions and effects of feedback in technology-based learning environments. The first two chapters detail what a technology-based learning environment is, the factors to consider and suggestions for designing an effective one. This chapter serves to present major theoretical perspectives and research findings on feedback in all types of learning environments. Specifically, in the following sections, the definitions, types and theoretical perspective of feedback will be described first. Then, past and more current research findings concerning feedback research will be discussed.

Definitions and Types of Feedback

Feedback, according to Webster's New World Dictionary (1976), refers to "a process in which the factors that produce a result are themselves modified, corrected, strengthened, etc. by that result" (p. 513). Thus, it is commonly believed that feedback, when employed in learning situations, should contain corrective or evaluative information which follows a learner's response. Often times, feedback is applied as a single instructional unit and its form can include the simplest "yes" or "no" message, to the indication of the correct answer, and eventually, to the presentation of corrective information or remedial instruction (Carter, 1984; Clariana & Lee, 2001; Kulhavy, 1977; Lee, 1994; Lee & Dwyer, 1994; Lee, Savenye, & Smith, 1991).

Thus, in addition to having defined feedback as a single, generic corrective agent, most researchers classified feedback into various types based on the amount and complexity of information contained in a feedback message. Most of these researchers believe that the level of complexity, or the amount of information in feedback, serves different functions in learning. For example, Carter (1984) reported that feedback is usually constructed into three to four forms. These forms are: (a) knowledge of response (KOR) feedback, which contains simple messages stating "correct", or "incorrect" and this type of feedback is used simply to let the learner know the result of his/her response to a question; (b) knowledge of correct answer (KCR) feedback, which informs and helps learners as to what the correct answer is; (c) corrective feedback, which states why an incorrect response is erroneous and why a correct response is correct; or (d) various embellishments of these three basic types. The more elaborative type of feedback often contains more information with a corrective or remedial nature with the hope that the learner will use the feedback information to correct a previous error. Similarly, Dempsey, Driscoll, and Swindell (1993a, 1993b) came up with five different kinds of feedback: (1) No feedback, which gives literally no feedback at all but still requires the learner to answer or respond to a question; (2) Simple verification feedback, which merely states whether the learner's response is correct or incorrect; (3) Correct response feedback, which states what the correct answer is; (4) Elaborated feedback, which affirms whether the answer was correct or not and explains why the answer was correct or incorrect; and (5) Try again feedback, which states whether the answer or response was correct or not but allows one more try.

In addition, Mason and Bruning (2001) listed the following five types of feedback : (1) no feedback; (2) knowledge-of-response feedback; (3) answer-until-correct feedback; (4) knowledge-of-correct-response feedback; (5) topic-contingent feedback; (5) response-contingent feedback; (6) bug-related feedback; and (7) attribute-isolation feedback. In essence,

Mason and Bruning's definitions of all these seven types of feedback are very close to the five basic types of feedback that are covered in this chapter and used by most researchers. However, except for knowledge-of-response and knowledge-of-correct response feedback, the rest of the terms explained by Mason and Bruning were somewhat different from the common feedback types described by other researchers. For example, answer-until-correct feedback is explained as a modification of knowledge-of-response feedback that is derived from master learning instruction, which provides verification without elaboration and requires the learner to stay with the same missed question until a correct response is produced. Topic-contingent feedback offers the learner verification of a correct or incorrect response plus elaborative content concerning the targeted topic. With this feedback, following an error, the learner is required to review the missed content in the lesson and from which he or she may correct his/her misconceptions and produces a correct response. Response-contingent feedback is also called extra-instructional feedback. This type of feedback functions exactly like informational or explanatory feedback, which explains why an error is erroneous and why a correct response is correct. Bug-related feedback informs an error and addresses it. This feedback does not provide the learner with the correct answer but offer related information for self-correction. Attribute-isolation feedback offers verification plus key components of the concept in order to facilitate better understanding of the concept. All of these terms have not been commonly employed in the feedback literature.

Still, the three most commonly used, mentioned or researched feedback types are: (1) Knowledge of response (KOR) feedback, which simply informs the learner if his or her response or answer to a question was correct or incorrect. KOR feedback could contain messages ranging from, for example, "Right", "Correct", and "You Got It!" (2) Knowledge of the correct answer (KCR) feedback, which is provided following both a correct and an incorrect response and simply provides the correct answer to a question. An example of a KCR feedback would be "The Correct answer is…" and (3) Explanatory, corrective or more elaborative type of feedback, which provides explanations for why a correct response is correct and why an incorrect response is erroneous and how to correct an erroneous response. In studies of Spock (1988) and Lee (1989), a sample question, the correct answer of the question and the corrective or explanatory type of feedback was presented as follows:

Which of the following statements describes the function of variables?

(1) Variables represent values in memory, values which cannot be changed.

(2) Variables represent values which can be changed but are not always stored.

(3) Variables represent values in memory, values which can always be changed (this is the correct answer).

(4) Variables represent an initial value which can not be changed later.

Explanatory feedback for question 1 when response 1 is selected:

No, values of variables CAN change. The whole reason for having variables is that their value can vary depending on what is entered into the computer. The correct answer is number 3. Each variable reserves a place in memory where its value is stored and changed, just as a house address indicates a location whose occupants may change.

Explanatory feedback for question 1 when response 2 is selected:

Sorry, number 2 is incorrect. Values of variables can be changed, but these values are always stored at an address in memory. Each variable reserves its own place with an address in memory where its value is stored and changed, just as a house address indicates a location whose occupants may change. The correct answer is number 3.

Explanatory feedback for question 1 when response 4 is selected:

Sorry, number 4 is incorrect. The address where the variable's value is stored is an address in memory. A variable must be stored in memory to use it, just as an idea must be in your own memory for you to recall it and think about it. The correct answer is number 3. Variables are stored in memory so that their values may be changed as needed.

Explanatory feedback for question 1 when response 5 is selected:

Sorry, number 5 is incorrect because the initial value of a variable can later be varied. The correct answer is number 3. The whole idea of having a variable is to have a symbol represent a value that can be changed, depending upon what's entered into the computer. Each variable has its own address in memory just as a house address indicates a location whose occupants may change.

Feedback Functions: Evolving from a Behaviorist To a Cognitive Point of View

In the past, operant, behaviorist psychologists regarded that feedback simply serves as a reward for reinforcing or repeating a correct response. Thorndike examined the use of feedback as early as 1911. He soon came up with the Law of Effect, which explains that feedback is expected to act as a connector between responses and preceding stimuli (as cited in

Mory, 1996; Kulhavy, & Wager, 1993). According to Thorndike, a response that was followed by feedback stating the response was correct is most likely to be repeated. An often-employed device to demonstrate the effect of rewards on the animal's response was the Skinner box. In this box, the animal's correct responses were immediately rewarded. After a short while, the animal consistently performed the desired responses. Skinner (1951, 1957) therefore maintained that once the animal's correct response was reinforced, the probability of its reoccurrence increased. As for human learners, Skinner (1968) believed that immediate feedback given after every correct response was necessary in learning because it functioned "not only to shape the behavior most efficiently but to maintain it in strength" (p. 39). Based on such a perspective, in the past, many researchers have explained the functions of feedback in terms of Skinner's reinforcement theory. As a result, when those researchers designed learning programs following operant reinforcement theory, they often reinforced the learner for every correct response by means of immediate feedback.

However, many other researchers disagreed with Skinner regarding the function of feedback (Ammons, 1956; Cohen, 1985; Kulhavy & Anderson, 1972, 1977; Kulhavy, Yekovich, & Dyer, 1979; Kulhavy & Wager, 1993; Lee, 1989, 1991, 1993, 1994; Lee & Dwyer, 1994; Smith & Smith, 1966). For example, as early as 1956, Ammons questioned if feedback simply acts in a reinforcing manner. He reviewed the literature on feedback and found that learners who were given the amount and direction of their errors performed better than those who were given knowledge of correct response, as prescribed by Skinner (1951, 1957). As described, the Skinnerian approach proposed that continuous reinforcement was most effective for increasing the repetition of a correct response. Thus, in the situation of an error, the Skinnerian approach cannot be applied so as to provide the learner with enough information to correct himself or herself in favor of the correct answer. With respect to errors, Ammons made the point that poor performance was often a result of learners' incorrect hypotheses about how to make correct responses. In Ammons's opinion, learners usually have hypotheses about what they are to do and how they are to do it while answering a question. When learners have false hypotheses and are unable to change those hypotheses, they perform poorly. In such a situation, with limited knowledge of results, as proposed by Skinner's theory, learners are not prepared to correct their false hypotheses. Therefore, Ammons concluded that feedback should contain specific and corrective information which can be utilized by the learner to correct false hypotheses. He, therefore, maintained, "The more specific the knowledge of performance, the more rapid the improvement and the higher level of performance" (p. 287).

Cybernetic psychologists including Smith and Smith (1966), who have compared the human control mechanism with electro-mechanical control systems such as computers, agree with Ammons and have attempted to define feedback from their perspective. First of all, Smith and Smith pointed out that feedback could not be appropriated as a form of reinforcement unless learners always make correct responses. According to these researchers, if the feedback signal only contains a reinforcing property, then "the smaller the magnitude of the error, the greater the reinforcement value of the signal" (p. 204). Undoubtedly, cybernetic theorists disagreed with Skinner that learning simply involves establishing a reinforcement schedule so as to strengthen correct responses. Smith and Smith further pointed out that feedback should primarily function to detect errors and to give information to redirect learning. Within the cybernetic context, the individual learner is viewed as a feedback system which generates his/her own activities in order to detect as well as control specific stimulus characteristics of the learning situation. Specifically, Smith and Smith maintained that, in a feedback-control system, a learner is described to have three primary functions: "it generates movement of the system toward a target or in a defined path; it compares the effects of its action with the true path and detects error; and it utilizes this error signal to redirect the system" (p. 203). Thus, the term feedback is used to portray a kind of reciprocal interaction between a series of activities. Among these activities, one activity generates a successive action which in turn redirects the primary action. As a result, cybernetic theorists asserted that learning is determined by the nature of the feedback-control processes available to the individual learner. And, the learning situation should be constructed in a way that fits the control capabilities of the learner. Therefore, the role of feedback is mainly for aiding the learner to generate a course of action and, when necessary, to redirect or correct that action through using the information provided.

In addition, many other cognitive theorists agree with Ammons (1956) and Smith and Smith (1966) and pointed out that feedback is not used solely for reinforcing overt, correct behaviors. Instead, feedback is assumed to be corrective and given in order to facilitate learners' internal processing of learning as well as aid them in correcting misconceptions (Gagne, 1970, 1977; Phye, 1979; Sternberg, 1983). Broadly speaking, a cognitive learning view is about how learning is a product achieving from an internal, cognitively monitored process to the content received. Therefore, with this approach, making or producing a particular response is "an internal process, as is also the decision to emit them on a particular occasion and under particular condition" (Caroll, 1976, p. 17). Specifically, cognitive theorists view the learner as an

45

information processing system and focused their research on how a learner acquire, rehearse, retain and recall the newly learned information among three levels of human memory storage: sensory memory, short-term memory and long-term memory (Gagne, 1970, 1977; Gagne, Briggs, & Wager, 1987; Gagne & Medsker, 1996; Gagne & White, 1978).

Further, Gagne (1977) also reported that learning that occurs in a processing model can be improved whenever necessary through feedback received. This is because the entire information-processing procedure can be checked or modified with feedback information provided. Gagne, thus, stressed that "Although feedback usually requires an external check, its major effects are obviously internal ones that serve to fix learning, to make it permanently available" (p. 75). Specifically, as Phye (1979) pointed out, feedback serves to fix learning in two ways: when a correct response is produced, feedback is used to confirm that learning has achieved its purpose and this strengthens the association between the practicing task and the related knowledge structure in long-term memory. While an error occurs, feedback is used to provide the learner with meaningful and explanatory information for the modification of the existing information-processing process. The learner can use feedback information to further direct him/her to revise each stage as well as the entire flow of information processing process.

In 1968, Lublin asserted that feedback has less cognitive effect on a correct response as on an error. As a matter of fact, Lublin found that learners performed better when there was no feedback provided to confirm a correct response. Lublin believed that the omission of feedback after a correct response may demand learners to attend to the information more deeply so that they learned better. Lublin's belief could be supported by explanations from a cognitive perspective. When a correct response is confirmed by feedback following it, the learner may stop processing the relevant information simply because he/she knows that his/her response was correct. In contrast, when the learner is informed about an error, he/she has to develop appropriate cognitive strategies so as to determine why an error occurred, to discover the difference between the correct answer and his/her error, to clarify his/her misconceptions and to reason out which answer is correct and why. Such a process requires the learner to analyze, organize, synthesize as well as retrieve the information to a greater degree. Therefore, this process produces a better information processing process.

In addition, researchers such as Ausubel (1968) and Kulhavy (1977) had attempted to give specific definition as well as to describe, theoretically, how feedback works. Ausubel (1968) stated that the cognitive effect of feedback on learning lies in its confirming and corrective capabilities. With these capabilities, feedback can function to confirm appropriate meanings and associations, correct errors, clarify misconceptions, and indicate whether or not learning has achieved its purpose. Consequently, feedback can inform the learner if he/she has reached the goal or how close he/she has come to it. Therefore, the learner's confidence in the validity of his/her learning achievements is increased. Further, by studying the feedback information, the learner can be aided in the location of errors and the focusing of his/her efforts and attention on those aspects of the task requiring further refinement. Thus, the cognitive aspects of feedback can increase the stability, clarity and discriminability of meaningfully-learned ideas.

Kulhavy (1977) further pointed out that feedback can be treated as a unitary variable. With that, as stated earlier, Kulhavy explained that the form or composition of feedback can range along "a continuum from the simplest 'Yes-No' format to the presentation of substantial corrective or remedial information that may extend the response content, or even add new material to it" (p. 212). In his opinion, the following assumptions must be met so that feedback can yield effects: first, the response should precede feedback. Responding first can prevent learners from peeking at the correct answer. Second, the correct answer should not be easily accessible. When the answer is not handy, learners are more likely to recall the learning materials than simply copy the answer from feedback message. Third, the instructional materials should be organized or presented in a way that learners are able to comprehend them or fit them into an existing knowledge structure. If the materials are too difficult or too unfamiliar to learners, learners may be unable to comprehend the questions and then guess at answers. In such a situation, feedback may become useless. Further, Kulhavy stated, when one considers specifically how feedback facilitates or promotes learning, one should consider the correct and incorrect responses separately. He, also, argued that feedback after an error is far more important than feedback provided following a correct response. As he pointed out, in the case of a correct response, feedback is utilized to tell the learner that his/her overall understanding of the material is accurate up to that point. Therefore, feedback following the correct answer simply serves to provide "a mechanism within the student-lesson system for informing the controlling unit (student) that the comprehension strategy is achieving the terminal goal of transferring the information from text to learner" (p. 220). On the other hand, when the learner is informed that he/she is in error, he/she begins to perceive that his/her comprehension or interpretation of the learning materials is inaccurate. He/she then has to replace his/her misconception with the corrective information in order to correct his/her mistake. Under such a condition, the information contained in feedback not only serves to help the learner identify an error, but also provides the learner with the needed information for self-correction purposes. As long as the learner is given substantial or remedial feedback information, he/she is equipped with more knowledge and direction in searching for the correct answer.

In summary, traditional operant, behaviorist psychologists regarded feedback as a reward for reinforcing the correct response. Skinner (1968) believed that immediate feedback provided after every correct response could maintain that response in strength. Many psychologists have disagreed on appropriating feedback as a form of reinforcement. They maintained that problems existed in applying the Skinnerian approach when errors occur. Ammons (1956) argued that feedback primarily functions to correct errors, rather than to reinforce correct responses. He also maintained that feedback should contain corrective information which can be used by the learner to correct his or her false hypotheses about the answer. Cybernetic theorists (Smith & Smith, 1966) believe that feedback should function to direct learners to generate a course of action and then to detect errors or to correct that action. Gagne (1977) further asserted that feedback is not used solely for reinforcing overt behaviors. Rather, feedback is assumed to be corrective and its major function is to facilitate learners' internal processing of learning. Ausubel (1968) and Kulhavy (1977) both urged that feedback after an error is far more important than feedback after a correct response. In Kulhavy's opinion, feedback after a correct response simply serves to tell the learner that his or her understanding of the material is correct. In contrast, feedback after an error not only helps the learner to identify the error but also provides the learner with the information necessary for self-correction. Once the learner is aided by the corrective information provided, he or she can begin the journey of clarifying misconceptions and determining the correct answer.

Feedback Functions: from a Constructivist View of Learning

With constructivism, an alternative view about how learners construct their own experiences to learn becomes available. The view of constructivism is that learning via activities in realistic learning contexts is the best way for learners to learn. Most constructivists argue that traditional instructional approaches over-simplify most real-world knowledge with the inherent result that learning is shallow and not long lasting. Constructivists believe that learning is a process; therefore, learning is best manifested in open-ended, real-world experiences. Flexible approaches to learning including allowing learners to experience and construct their own learning, and making learning to be embedded in realistic contexts should be valued. Furthermore, constructivists advocate the opinion that learning is enhanced and buffered by social interaction. Multiple representations of concepts being learned are essential because these add realism. Learners need to be acutely aware of their learning process and frequently encouraged to reflect on growth, goals, and needs. Constructivism also has applications in the field of technology-based learning with a view that the most effective way to gain knowledge and learning is from active dialogue and engagement among those who seek to understand and apply concepts and techniques (Duffy & Jonassen, 1992; Jonassen, 1996, 2006).

Accordingly, in constructivism, as explained by Mory (1996), feedback must fulfill several needs. First, feedback should guide the learner towards his or her personal reality. Second, effective feedback in a constructivist program must aid in building symbols that hold meaning for the learner. Third, the feedback provided must be placed in context with real-world experiences. Feedback must also derive meaning that can be determined by internal understanding. Lastly, feedback should be similar to a construction tool for the mind. Transmission of feedback in a constructivist learning environment can be achieved through social interaction with other learners. This allows a system of correction through interaction. With respect to the use of feedback, there are three main constructivist principles that are currently applied: situated cognition, cognitive flexibility, and microworlds. Situated cognition rests on the belief that instruction should be delivered in a relevant context (Brown, Collins, & Duguid, 1989) and that the context will become part of the knowledge base for learning. This theory can be applied by using cognitive apprenticeship, which is when learners engage in activity and utilize both physical and social perspectives, like an apprentice. Feedback would be attained through one's peers as learners solved problems in a group, team-oriented manner. Instead of acquiring predetermined instructional sequences, feedback could act like a coaching mechanism that scrutinizes strategies used to solve problems. Cognitive flexibility theory, on the other hand, involves presentation of different contexts to viewers. Spiro, Feltovich, Jacobson, and Coulson (1991) proposed using multidimensional hypertext to convey content, and feedback can be implemented on a hypertext system to lead the learner to approach concepts from different perspectives. Via focusing on conceptual interrelation, giving many different perspectives to look at problems from, and emphasizing "real-world instruction", feedback can help learners gain advanced knowledge in learning domains that are ill-structured with no single correct answer. The theory stresses being able to look at problems from all perspectives. For instance, if a learner is approaching a problem from one perspective and not succeeding, feedback can be used to suggest looking at the problem through a different or additional context. Lastly, the microworld principle is based on a subset with a small scale of reality in which the learner can obtain information about the object based on his/her choice. Most microworld environments, Rieber (1992) contends, should be designed with a feedback loop that provides a continual stream of information that aids students in establishing and maintaining goal setting and goal monitoring. In addition, microworlds can set the learning environment into a finite set of variables. Any received feedback can be judged and compared against a learner's individually designated goals. With a microworlds system, multiple kinds of feedback are sought. Overall, feedback in a constructivist environment should achieve several goals. It must aid the learner in constructing an internal reality by providing intellectual tools. It also should help the learner solve problems within relevant settings. In addition, it must challenge the learner into developing their own perception of reality or the context of learning. Lastly, it should guide the learner through constructing of his or her own reality, experiences as well as learning (Mory, 1996).

Overview of Previous Research on Feedback

As stated, the functions and effects of feedback were once explained and investigated from a behaviorist point of view. Again, based on a behaviorist point of view, positive reinforces are used to stimulate the reoccurrence of a desirable, observable behavior. According, based on this view, feedback should be provided immediately following every correct response so that such a desirable response will repeat itself again. Thus, previous studies on feedback functions and effects were centered on the issues concerning (1) whether it is the reinforcing or informational characteristic of feedback that facilitates learning; (2) whether it is the immediate or the delayed feedback that enhance learning most; and (3) whether the level of information provided by feedback produces different learning results. The following sections details research findings in each of these issues.

Feedback Effects: Reinforcing or Informational

As early as 1956, Buss and Buss (1956) investigated the effect of verbal reinforcement combinations on conceptual learning. The specific reinforcement combination in their study was: (a) Right for a correct response, Wrong for an incorrect response (Right-Wrong); (b) nothing for a correct response, Wrong for an incorrect response (Nothing-Wrong); (c) Right for a correct response, Nothing for an incorrect response (Right-Nothing). They found that the subjects who merely received the message of "Right" following their correct responses (nothing following their incorrect responses) performed worst. Also, the subjects who received "Right" for their correct responses and "Wrong" for their incorrect responses performed similarly to those who received nothing for their correct responses and "Wrong" for their incorrect responses. Such findings were contrary to the proposition in most behaviorist view that a correct response will repeat itself and therefore learning will be boosted if knowledge of correct response is provided immediately after every correct response.

Sassenrath (1975) pursued the question of the function of feedback by analyzing three previous studies by Sassenrath (1972), Sassenrath and Yonge, (1968, 1969) to investigate if feedback given immediately after the correct response would maintain that response in strength. Tasks involved in those studies were verbal learning. Generally, subjects had to learn to associate the correct word or phrase with a series of words or sentences. Either immediate or delayed feedback was offered

53

following the responses. After analyzing the data from those studies, two types of ratio scores were obtained: (a) the change from a wrong response on first test (WI) to a right response on the second test (R2) and (b) the repetition from a right response on the first test (Rl) to a right response on the second test (R2). Primarily, data were rescored for the ratio of subjects repeating their correct responses between the first and second test, R2/Rl. If feedback were reinforcing, immediate feedback would produce a higher R2/Rl score than delayed feedback. The results showed that immediate feedback did not produce a higher R2/Rl score than delayed feedback. Actually, delayed feedback was superior in helping students change their incorrect responses on the first test to correct responses on the second test, Wl/R2. These findings indicated that immediate feedback does not work to increase the ratio of subjects' initial correct responses. Rather, feedback yielded more effects on correcting initial errors and making more correct responses on later tests.

Bardwell (1981) also questioned if feedback could be explained in terms of the reinforcement theory. In her study, the function of feedback was studied in a reinforcement theory sense by comparing immediate feedback to delayed feedback. Subjects involved in her study were 4th, 6th and 8th graders and were equally divided between immediate and delayed feedback groups. All subjects learned German words. Two immediate recognition tests, a posttest and a delayed retention test were administered to measure learning. Under the immediate-feedback condition, subjects' correct responses were immediately marked while they were taking the recognition test. Under the delayed-feedback condition, subjects took the recognition test without receiving feedback. Then they were allowed to study the corrected test paper in the following day. As Bardwell pointed out, behaviorists often argue that if feedback is immediate, only the overt response will be reinforced. In other words, in the situation of learning foreign words, behaviorists would predict that immediate feedback would facilitate learning speed of language acquisition. On the other hand, such theorists would also expect that delayed feedback would facilitate retention. The reason for such expectation is whatever learned under the condition of delayed feedback would be more resistant to extinction (p. 4). Bardwell thus suggested that feedback would be akin to reinforcement if delayed feedback facilitated retention but hindered acquisition. On the contrary, feedback would be informational if delayed feedback facilitated both acquisition and retention. The research results showed that a main effect favoring delayed feedback for both acquisition and retention. Bardwell, therefore, concluded that feedback serves as informational event and does not work in a reinforcing manner.

In addition, Moore and Smith (1964) conducted studies to investigate whether providing statements of praise or extrinsic reward such as money after the correct response would boost learning. In More and Smith's study, college students were required to perform a verbal information task concerning Skinner's (1951, 1957) theory. During practice, they were given: no knowledge of results; (b) immediate knowledge of what the correct answer should be; (c) immediate knowledge about the correctness of the response (a flashing light for correct responses) and (d) immediate knowledge about the correctness of the response plus one penny for each correct response. The results showed that these different feedback conditions produced no differential effect on student achievement.

To summarize, all of these findings indicate that the functioning of feedback in instruction is informational, rather than reinforcing. If feedback were reinforcing, providing immediate feedback to confirm correct responses would work better than delayed feedback. In other words, if feedback functioned solely as a reinforcing agent, immediate feedback would have a greater facilitative effect on learning than delayed feedback. As described, Bardwell (1981) found that immediate feedback did not yield a greater facilitative effect on acquisition and retention than did delayed feedback. Buss and Buss (1956), More and Smith (1965) and Sassenrath (1975) all revealed that confirmational feedback did not promote learning or increase the repetition of an initial correct response. Rather, they all found that telling students when they were wrong resulted in better performance. It is the informational aspect of feedback that allows a learner to locate errors and to correct them. As long as errors are eliminated or have been replaced by correct information, the learner's understanding of the learning material is, once again, accurate, therefore increasing what he or she gains from instruction.

Timing of Feedback

While there is increasing evidence that feedback is informational and facilitative rather than acting solely as reinforcement, there has developed a controversy regarding whether feedback should be immediate or delayed. Again, according to Skinner (1951, 1957, 1968), feedback should follow a response as quickly as possible. As mentioned, for years proponents of programmed instruction and teaching machines emphasized the principle of immediate feedback following a correct stimulus-response bond. However, many researchers questioned whether or not this principle could be generalized to school learning. They argued that there is a relationship between adequate timing of feedback and other variables such as learning tasks and the level of mastery of the learner (Carter, 1984; Cohen, 1985; Kulhavy & Anderson, 1972; Kulik & Kulik, 1988; Rankin & Trepper, 1978).

With respect to the level of mastery of the learner, Cohen (1985) maintained that for low-mastery learners, immediate feedback is necessary because it provides constant help to facilitate their proceeding through the learning content. For high-mastery learners, immediate feedback may become redundant to them and thus, may impede the pace of learning. Cohen (1985) also suggested that, the delayed feedback which is given after the entire unit of items is best for high-mastery students, because such feedback allows them to assess how they perform on each item in the unit as well as to internally process the content without impeding its flow. Furthermore, Anderson, Conrad, and Corbett (1989) conducted a study in which subjects were trained using the same program but with different kinds of feedback. The study showed that students who received immediate feedback went through the program faster, but did not do any better than the students who received delayed feedback.

Carter (1984) suggested that when the learning task requires high level of learning such as learning and retaining new knowledge, feedback should be delayed. When the learning task is discriminating in nature or of a lower cognitive level, the timing of feedback should be immediate. A possible reason for such suggestions may be that during discrimination or low cognitive level of learning, immediate feedback can constantly provide remedial information to help learners acquire knowledge as well as work out confusion (Siegel & Misselt, 1984). In contrast, during high level of learning, delayed feedback offers evaluative and corrective information so as to work as a summary and internal organizer, which facilitates the learning of conceptual and abstract information (Cohen, 1985).

Various studies have been conducted to investigate timing of feedback in order to determine when to provide feedback. Among these studies, immediate feedback is defined as that which is given with 0 second delay after every response. Three kinds of delayed feedback often investigated were: (a) feedback was given item-by-item with a pre-established number of seconds delay; (b) feedback was given after an entire sequence of items with 0-second delay and (c) feedback was given after an entire sequence of items with an established number of minutes, hours or days delay.

More (1969) examined the effects of four different delayed feedback schedules ranging from delay by 0 second, by a few hours and days. Eighth graders were asked to perform a verbal learning task regarding learning science and social studies information. More found that the group who received no delay feedback scored significantly lower than the other three delayed feedback groups in immediate acquisition test. The groups who received a few hours of delayed feedback yielded significantly higher scores on a delayed retention test than did the 0-second and 4-day delayed feedback groups. Such findings have indicated that delay of feedback may improve long-term retention and the delay which predicts optimal retention was about 1 day. Rankin and Trepper (1978) and Gaynor (1981) all investigated the effects of (a) feedback given item-by-item with either no or some delay and (b) feedback given immediately after the entire sequence of items (end-of-session feedback). Rankin and Trepper found that in a computer-assisted lesson that dealt with a high cognitive task regarding learning meaningful materials on sex education, feedback delayed for 15 seconds and feedback given after the entire section of items produced better retention effect than did immediate feedback. Gaynor looked at Bloom's (1956) Taxonomy to examine how timing of feedback affected different taxonomic levels for students at low and high mastery levels. She found that immediate item-by-item knowledge of correct response feedback was beneficial for students at low mastery level on initial acquisition of knowledge. And, end-of-session feedback which showed subjects' progress in the unit facilitated learning at the comprehension and application level for high-mastery students. She also reported that feedback delayed for 30 seconds after every response had a detrimental effect on retention at the knowledge acquisition and application level. However, such feedback did not yield a detrimental effect at the comprehension level for those who showed high mastery of the task. These findings were consistent with both Cohen (1985) and Carter's (1984) view that immediate feedback can provide constant help to low-mastery students in acquiring information and working out confusion. And delayed feedback is beneficial to high-mastery learners in assessing their performance and consolidating conceptual materials.

Also, Kulhavy and Anderson (1972) and Surber and Anderson (1975) investigated the effects of immediate and delayed feedback on the probability of repeating an initial error on a later test. In Kulhavy's study, high school students completed a multiple-choice test on topics in introductory psychology and received knowledge of the correct answer feedback with 0-second, one or two day delay. Subjects then took a delayed test seven days later after the experiment. The results indicated that the probability of repeating an initial error on the delayed test was significantly lower for delayed feedback groups than for immediate feedback group. Accordingly, Suber and Anderson used high school students to perform a higher cognitive task of studying a meaningful passage on army ants. Either no feedback, immediate feedback or delayed feedback was administered. They found that when the measure was the probability of changing an initial error to a correct response on a later test, feedback proved superior to no feedback, and delayed feedback proved significantly better than immediate feedback. As these researchers explained, such results are due to the fact that learners forget their errors over the delayed interval, and thus there is less interference with learning the correct answer from feedback. Therefore, learners become capable of changing an initial error to a correct response on a later test. It should be noted that this task was a verbal information learning task, rather than an intellectual skill.

In summary, four studies reviewed provide evidence that there is a relationship between timing of feedback, learning tasks and learners' level of mastery. It appears that delayed feedback facilitates learning better than immediate feedback when learning involves meaningful verbal learning and retention. In contrast, immediate feedback facilitates learning better than delayed feedback when learning involves initial learning tasks such as discrimination learning and concept formation. As for learners at higher mastery levels, delayed feedback showed more effects in aiding learners to review the entire procedure of information processing without impeding its flow. For learners at lower mastery levels, immediate feedback is more useful in providing constant help in order to facilitate learners' proceeding through the learning content (Cohen, 1985; Gaynor, 1981; Rankin & Trepper, 1978). On the other hand, instead of providing learners with feedback during practice, Kulhavy et al. (1972) found that delayed feedback was more effective than immediate feedback in reducing the probability of repeating an initial error on a later test. Also, delayed feedback was more effective than immediate feedback in changing an initial error to a correct response on a later test.

Schedules of Feedback

In addition to investigating timing of feedback, previous research was also concerned about schedules of feedback. According to Carter (1984), scheduling of feedback relates to the decision as to whether feedback should be provided subsequent to correct responses, incorrect responses, a specified proportion of correct or incorrect responses, or subsequent to all responses. Theoretically, he claimed that if the learning task is verbal discrimination in nature, which usually requires learners to contrast the differences between the correct word and a number of incorrect words, feedback should be provided subsequent to incorrect responses and should contain error identification information. The reason for such claim may be that during discrimination learning, simple knowledge of success or failure can not aid learners to work out confusion or reason out why a correct answer is correct. Rather, when the discrimination errors occur, learners often need information about the direction and amount of their errors in order to correct them.

The effects of feedback schedules on learning were investigated by Lublin (1965) and Anderson, Kulhavy and Andre (1971). In Lublin's experiment, college students were asked to perform a high level cognitive task of learning the concept of Holland and Skinner's Analysis of Behavior. Different groups of subjects received (a) no feedback, where all knowledge of the correct answers were not provided; (b) continuous knowledge of the correct answer (KCR) feedback; (c) variable-ratio 50% KCR feedback, where a random 50% of the frames were followed by the correct answer and the remaining 50% of the frames were not followed by the correct answer; (d) fixed-ratio 50% KCR feedback, where frames with even numbers were followed by the correct answer and frames with odd numbers were not followed by the correct answer. Lublin found that on the posttest test, the no feedback group scored highest, followed by the variable-ratio 50% group, followed by the fixed-ratio 50% group. The continuous KCR feedback group scored worst on the test. Although such results have indicated that different schedules of feedback can yield differential effects on learning, they are contrary to most research results that feedback promotes learning better than no feedback. Lublin gave a possible explanation for these conflicting findings: The removal of confirmational feedback for correct responses required more attention from the subjects and forced them to carefully study each frame; therefore the no feedback group performed best. In other words, in Lublin's opinion, students who received no feedback following their responses may have been more careful about formulating and testing their hypotheses about the answer and, thus, may have made fewer errors. In addition, subjects who received no feedback may have demanded themselves to process the instruction to a deeper degree so that their performance improved.

Accordingly, Anderson, Kulhavy and Andre (1971) used college subjects in two experiments to complete a computer-assisted lesson with a higher cognitive learning task (diagnosis of myocardial infection). In the first experiment, they compared the effects of eight different feedback schedules: (a) no knowledge of the correct answer; (b) KCR following every item; (c) KCR following correct responses only; (d) KCR following a random 10% of correct responses; (e) KCR following errors only; (f) KCR following errors with a 15-second time out (the frame with the missed item was reviewed for 15 seconds prior to KCR); (g) KCR following correct responses and KCR following ten errors on the same missed item and (h) KCR or no KCR according to the subject's choice. The results showed that groups with KCR following either every item, correct responses only, errors with time out or the learner's choice performed significantly better than the group with no KCR. Different schedules of feedback did not reveal differential effects upon learning achievement. In the second experiment, in addition to feedback conditions with no KCR, the item-by-item KCR and KCR with timeout, three other conditions were also administered: (a) subjects who received KCR after every response were forced to try out the missed item until they responded correctly; (b) subjects who received KCR after every response were provided the missed item according to a specific review schedule and (c) subjects could access KCR prior to giving responses. The results revealed again that in a computer-assisted course with a high cognitive learning task, different schedules of feedback were not differentially related to achievement. In addition, the item-by-item KCR group learned more than the no KCR or KCR prior to responses groups. The effects of other feedback conditions in this study were not significant. Although neither of these two experiments confirmed that feedback scheduling is critical to learning, they all showed that learners receiving constant KCR feedback learned better than those who received no feedback.

Types of Feedback

Some researchers preferred to treat feedback as a unitary instructional variable and did not classify it into various categories. However, many researchers have examined the effectiveness of different types or amount of feedback information so as to determine how much information and what type of information should be contained in feedback. For example, Gilman (1969) investigated the effects of five different types of feedback in a computer-assisted lesson dealing with commonly misunderstood science concepts. College students were given

(a) no feedback; (b) KOR feedback; (c) KCR feedback; (d) corrective feedback with explanation regarding why a correct response was correct or why an incorrect response was erroneous and (e) a combination of KOR, KCR, and corrective feedback. The results revealed that KCR feedback, corrective feedback, or a combination of KOR, KCR and corrective feedback produced better effects upon immediate retention than did no feedback or KOR feedback. And, subjects who received the most detailed feedback information outperformed other groups in terms of producing a higher number of correct responses in the posttest. These findings indicated that the amount of information that subjects received from feedback was critical to learning.

In addition, in Siegel and Misselt's (1984) study, the effects of corrective feedback, feedback schedules, and the immediate review of a missed item were investigated. In their study, college seniors engaged in a lower cognitive task consisting of learning 20 English-Japanese word pairs. Feedback was provided subsequent to incorrect responses only. Three kinds of feedback administered were: (a) knowledge of the correct answer feedback; (b) adaptive feedback, which informed the subjects regarding what stimulus they responded to and the answer to the stimulus and (c) adaptive feedback with discrimination training, which simultaneously informed the subjects about both the item missed and the item with which it was confused by the subjects. The missed item was reviewed immediately or repeated according to a spaced review schedule. The results indicated that subjects receiving adaptive feedback with discrimination training made significantly fewer discrimination errors than those who received adaptive feedback or knowledge of the correct answer feedback. Also, subjects receiving increasing ratio review made significantly fewer discrimination errors and out-of-list errors than those who received immediate review only. These findings confirmed Carter's (1984) view that the error correction information and scheduling of feedback were crucial when the learning task is discrimination.

In addition, Chanond (1988), using a computer-assisted lesson, gave several different kinds of feedback, and the study showed that immediate feedback that includes knowledge of correct answer, knowledge of why a subject's response was incorrect after an incorrect response greatly helped retention on both an immediate and a delayed posttest. Mory also performed a study in 1991, and varying combinations of task-specific feedback were used according to an assessment of answer correctness and a response certainty measurement. When compared to non-adaptive feedback, however, there were no large differences between these two conditions. Further, research results that favored feedback with increased information were revealed by Roper (1977). In Roper's study, college students completed a computer-assisted lesson on basic statistical concepts. The three feedback modes compared were (a) no feedback; (b) KOR feedback and (c) KOR feedback plus the correct answer stated in the context of the question. Roper found that among the three treatment groups, the group with KOR feedback in combination with the correct answer scored best on the posttest. No feedback was the least effective feedback condition. He also found that subjects who received KOR and the correct answer performed better in correcting previous mistakes on the posttest than subjects in other groups. Such findings confirm that the primary function of feedback is to aid the learner in locating errors and to provide information so that the learner can correct learning misconceptions or errors.

In addition, Roberts and Park (1984) examined the interaction between different amount of feedback information and learners' cognitive styles. In their study, learners were classified as field dependent or field independent and were asked to perform a higher cognitive task dealing with concept learning in psychology. During practice with a computer-assisted lesson, subjects received either KOR or explanatory feedback, which indicated whether the answer was correct or incorrect and provided information explaining the correct answer. First of all, Roberts and Park found that subjects who received explanatory feedback performed better on both the immediate posttest and the delayed retention test than subjects who received knowledge of response only, regardless of their cognitive style. They thus concluded that feedback that contains knowledge of result and explanatory information can benefit learners with both field dependent and field independent learning styles. Although these studies had revealed that feedback with additional explanatory information could better facilitate learning, there were other studies that did not show similar results. For example, in Hodes' (1984) study, ninth and tenth graders were asked to perform a higher cognitive task of learning metric conversions. Subjects used a computer-assisted lesson and received corrective or non-corrective feedback during practice. Corrective feedback informed an incorrect response as wrong, gave a corrective opportunity on the missed question and maintained supportive encouragement to try the missed question again. Non-corrective feedback informed an incorrect response as wrong, gave no corrective opportunity on the missed question, and gave no encouragement to try the missed question again. The results showed that corrective and non-corrective feedback did not yield differential effects upon retention.

In summary, feedback literature indicates that the amount of feedback information is critical to learning. However, there is little consensus on precisely how much information should be contained in feedback in order to maximize learning. Studies on the amount of feedback information have resulted in conflicting findings. For instance, Phye (1979) tested the prediction that the greatest amount of feedback information would yield the best immediate and delayed retention effects. In his study, college students were engaged in a lower cognitive task dealing with discrimination learning. After students took the pretest, they received (a) the minimal information feedback, which consisted of students pretest answer sheet with stems and correct answers only; (b) extensive information feedback, which consisted of the minimal informative feedback and the total number of correct responses on the pretest and (c) a combination of minimal and extensive informative feedback, which also consisted of individual items missed in the pretest. The results showed that the minimal informative feedback produced the greatest retention improvement on the posttest. According to Phye, a possible explanation for such a contradictory result may be that too much or too detailed feedback information seemed to confuse learners. When more than sufficient information was provided by feedback, it might distract learners' attention therefore may have a detrimental effect upon retention.

In conclusion, previous researchers found that feedback is not just for reinforcing correct answers. It can be used to help learner correct mistakes and produced better learning results. Some research found that feedback helped learners produce better learning results, but some research failed to reach such a conclusion. More detailed-oriented feedback research was conducted, such as research focusing on feedback types, timing and schedules. Some more conflicting findings were revealed with regard to levels of feedback. This may be due to the failure to control other variables such as placement of feedback, pretest, practice, posttest, learner's characteristics, learner's confidence level in a response, learner mastery levels, learning tasks, specific feedback timing or feedback schedules. For instance, two studies (Kulhavy & Anderson, 1972; Phye, 1979) discussed earlier, administered feedback during testing, rather than during practice. As described earlier, these studies asked learners to take tests and to receive feedback prior to their exposure to the content that those tests were designed to measure. Obviously, when learners were totally unfamiliar with the to-be-tested information, they could not use the test items as a retrieval cue to establish context for a memory search. Thus, they may have guessed at answers and the explanatory feedback may become useless. On the other hand, according to Mayer (1975, 1976), meaningful learning occurs when learners receive the to-be-learned information, have a meaningful learning set that can be used to assimilate new information and are able to activate the meaningful learning set during learning. Hence, in Kulhavy and Phye's studies, because learners were never exposed to the content, they can not have had or activated a meaningful learning set to which new feedback information could be assimilated. Under such conditions, learners who were given the detailed informative feedback may not have understood the feedback content and thus performed poorly on both the pretest and posttest. In contrast, learners who were given the simple KCR feedback may have tried to memorize the correct answer by rote during the pretest and were able to recall more correct responses on a posttest.

Due to these reasons, the focus of feedback has been switched to determine how and when feedback should be administered when considering all other learning or instructional variables simultaneously. Next section of this chapter covers more research studies on feedback that take additional learning or instructional variables into consideration.

Feedback with Other Variables

Clearly, feedback is a crucial component for learners to be engaged in a technology-based lesson. However, its effects and functions on learning have not been unequivocally determined. Behavioristically oriented theorists tend to appropriate feedback as a form of reinforcement. However, research results have more positive results on how feedback serves as an informational event and does not work in a reinforcing manner. Many researchers also questioned the superiority of immediate feedback. Studies on timing of feedback and schedules of feedback often obtain inconsistent results. It appears that for high-mastery learners, delayed-feedback is more effective than immediate feedback when the learning task is at higher cognitive level. On the contrary, for low-mastery learners with a lower cognitive task, immediate feedback is more effective than delayed feedback. Most studies on schedules of feedback were unable to show that different schedules of feedback have differential effect on learning. Only one study (Siegel & Misselt, 1984) reached the conclusion that increasing ratio review of the missed item could improve learning. With respect to feedback type, there is little consensus on what kind of feedback information or how much feedback information is most effective in facilitating learning. A number of studies on feedback type even showed that minimal feedback information resulted in best performance. However, these studies simply asked subjects to take tests and to receive feedback. Because subjects were totally unfamiliar with the testing content, they may have guessed at answers and feedback with extensive information may have become useless. Due to these reasons, some studies were conducted to determine, specifically the relationship between types of feedback, placement of feedback, and various other

69

variables. Other studies were also conducted to investigate whether the effectiveness of different types of feedback would vary with learners' characteristics, with the degrees of response confidence, with various learning tasks and with different levels of mastery of students.

For example, Roberts and Park (1984) examined the interaction between different amount of feedback information and learners' cognitive styles. They did not find any interaction effects between these variables. Another example, Kulhavy, Yekovich and Dyer (1979) tried to unveil if a relationship existed between the variables of different levels of feedback information provided and the degree of response confidence. These researchers used college undergraduates to study a program on heart disease and randomly assigned subjects to text review and no text review groups. While the subjects were responding to the questions, text review groups could review the instructional materials and no text review groups could not review those materials. At the same time, subjects had to determine the degree of confidence (ranging from low to high) in their responses to practice items before receiving either KCR feedback or no feedback. They found that subjects receiving KCR feedback made few errors and scored higher on the posttest. No interaction between feedback and text review was found. Yet, they found that no text review groups performed better than text review groups on the posttest. And, subjects were better able to correct a high-confidence error with KCR feedback than with no feedback. Thus, three conclusions drawn by these researchers were: (a) post response KCR feedback can benefit learners in correcting previous errors; (b) learners with no text review during responding performed better because the absence of text demanded a higher level of attention from them to study the learning materials and (c) an increased amount of the feedback information following high-confidence errors was beneficial because learners tended to study the feedback information more carefully under a condition in which they were sure that they were correct but found that they were wrong.

Spock (1988) also tried to investigate the interaction between KCR and explanatory feedback and the degree of response confidence. In his study, KCR feedback simply stated the correct answer and explanatory feedback consisted of stating the correct along with explanations regarding why the incorrect response was wrong and why the correct answer was right. Confidence of response was measured by asking subjects how confident they were about their response to a practice question was correct. Following a correct response, confident or not, KCR feedback was always given. In contrast, following an incorrect response, either KCR or explanatory feedback was provided for a given treatment depending upon the subject's confidence of response. The results showed that explanatory feedback was not superior to KCR feedback either in raising posttest scores or in improving learners' attitude toward feedback. However, the results did reveal tendencies toward interactions which indicated that learners' prior knowledge and confidence of response may influence the effectiveness of different types of feedback. According to Spock, one possible reason for those insignificant results may have been the difficulty of the instruction. He, therefore, recommended that future research is necessary on determining the relationship among difficulty of instruction, learners' prior knowledge and degree of confidence.

Finally, Lee (1989) extended the study of Spock (1988) to investigate the effects of three different types of feedback and their relationship with two additional variables: the learning task and an extra error correction opportunity. In her study, Lee provided a great amount of details and explanations regarding each of the hypotheses and findings of her study. Most of the details, information and comparisons of findings provided by Lee will be presented as follows, as the information could be insightful and may assist researchers to shed light on how the effects of different types of feedback intertwined with many other variables.

Specifically, in Lee's study, the focus was on revealing the effects of three types of feedback: KOR, KCR and explanatory feedback in combination with or without an extra error correction opportunity on error correction. Primary research questions addressed were: (a) what are the comparative benefits of KOR, KCR and explanatory feedback following an error during practice on posttest performance? (b) did an extra error correction opportunity as opposed to a single opportunity provided following an error during practice promote error correction on posttests? (c) was there an interaction effect between KOR, KCR and explanatory feedback in combination with number of error correction opportunities on posttest performance?

In Lee's (1989) study, for the main effects, the feedback factor, the research was conducted to reveal how different types of feedback worked on (a) increasing overall correct responses; (b) changing an initial error to a correct response on both the immediate and the delayed posttests, and (b) reporting a positive attitude toward the instruction. It was predicted that the explanatory feedback group would significantly outperform both the KOR and the KCR feedback group and the KCR feedback group would significantly outperform the KOR feedback group (explanatory feedback> KCR feedback> KOR feedback) with respect to (a) overall correct responses on both the immediate and delayed posttests, (b) the conditional probability of changing an initial practice error to a correct response on the two posttests. Regarding the conditional probability of repeating an initial practice error on the two posttests, explanatory feedback group would score significantly lower than the KOR or the KCR feedback group, and KCR feedback group would score significantly lower than KOR feedback group. In addition, it was predicted that subjects who received explanatory feedback would have a better attitude towards the instruction than those subjects who received either KOR or KCR feedback. In addition, subjects who received KCR feedback would like the instruction better than those who received KOR feedback. According to Lee, these effects were predicted because it was expected that compared to KCR and KOR feedback, explanatory feedback that included corrective information would better help the learner to correct an error during practice, thereby producing more correct responses on a later performance. And, since KCR feedback included more corrective information than did KOR feedback, it should better help the learner to determine why the correct answer was right and why the incorrect response was wrong during practice, thereby producing more correct responses on a later performance. Due to the above reasons, in a series of questions regarding the instruction (specifically attitude toward the feedback and number of tries), it was expected that learners who received explanatory feedback would score higher on those questions (higher score indicates more positive attitude toward the instruction) than those who received KCR or KOR feedback, and learners who received KCR feedback would score higher on those questions than those who received KOR feedback. The results revealed that the explanatory feedback group significantly outperformed the KOR feedback group. Although the means followed expected pattern (the explanatory feedback group > the KCR feedback > the KOR feedback), the differences between the groups of (a) explanatory and KCR feedback and (b) KCR and KOR feedback were not significant. In the delayed posttest, although means were still in predicted pattern (the explanatory feedback group > the KCR feedback > the KOR feedback), the differences between these three feedback groups were not significant. However, another measure, probability of repeating an error from practice on the delayed measure, did indicate a positive effect for level of feedback. Tukey's post hoc tests indicated that, as predicted, the explanatory feedback group was significantly superior to the KCR feedback group in reducing the probability of repeating a practice error on the delayed posttest. Although the explanatory feedback group performed better than the KOR feedback group in this same probability, the post hoc tests did not reveal a significant difference between these two groups. Similarly, although the KCR feedback group performed better than the KOR feedback group in this same probability, the post hoc tests did not reveal a significant difference between these two groups. This same effect was not found on the immediate posttest. Regarding the probability of changing a practice error to a correct response on a later performance, the results did not reveal a significant difference for the three feedback groups either on the immediate or delayed posttest. Neither did the results indicate any significant difference in attitude toward the instruction among the feedback groups.

In addition, Lee (1989) explained that although patterns of means were in the predicted direction on all measures examining effects of levels of feedback, the significance of these differences were inconsistently significant. Possible explanations for these effects are: (a) the learning motivation was suppressed by the difficulty of the learning materials and (b) low reliability of the testing instrument interjected more error variance. For example, the grand mean across all treatment groups was 7.24 for the immediate posttest, 6.25 for the delayed posttest, with total possible scores of 15 for each measure. So the average performance across all treatment groups was quite low. This indicated that the level of the difficulty and complexity of the instruction, practice questions and test items may have been beyond the ability level of most participating subjects. Therefore, the participating subjects may not have maintained the level of attention and motivation expected from them throughout the whole instruction. However, Lee explained that the instructional materials used in her study were intended to be written at a relatively difficult level in order to ensure some errors in practice so that feedback could be utilized.

With these findings, Lee (1989) provided further explanations regarding the findings of her study by citing the theory of Polya (1957), which proposed that a successful problem solving behavior requires understanding the problem, devising a plan, carrying out the plan and evaluating the effectiveness of the plan. Lee explained that in her study, when the participating subjects encountered difficulty in understanding or acquiring the learning information, they might not have been fully motivated to complete the whole lesson or to exercise any of the four problem solving stages in order to successfully answer the practice questions. If so, explanatory feedback given after a practice error may have worked as a remedial instruction. In such a case, if a test was administered immediately after the lesson, those who received explanatory feedback during practice may still have been able to recall the corrective information to produce the correct response on that test. In the same case, those who received KCR feedback may simply have memorized the correct answer, and matched that answer with an answer similar in form. As for learners in the KOR feedback group, if they were unmotivated or unable to learn, a simple message such as "You are right" or "You are wrong" may have been insufficient to aid them to analyze an error or to reason out the correct answer during practice, requiring more mental effort than they were able or willing to apply. This may have resulted in poor performance on the immediate posttest. These reasons may be responsible for the finding that a significant difference was only found in overall correct responses on the immediate posttest between the explanatory and the KOR feedback groups. These reasons may also be responsible for why the mean difference of the overall correct responses on the immediate posttest between the explanatory and the KCR feedback groups was in the predicted direction but not great enough to reach the statistical difference.

Also, in Lee's (1989) study, it was found that explanatory feedback was significantly better than KCR feedback in reducing the probability of repeating a practice error on the delayed posttest. Lee explained that the reasons for such a finding may be that if the participating subjects did not maintain a high level of attention or motivation throughout the entire computer-assisted lesson, they might not be able to employ their learning to answer the practice questions and might have missed most of the questions. If this were the case, the more practice errors, the more explanatory information would be received by the subjects in the explanatory feedback group. In the same case, a number of correct answers would be received by subjects in the KCR feedback group and the message containing fairly strong language such as "You are wrong" would be repeatedly received by subjects in the KOR feedback group. As a result, when the delayed posttest was administered, the subjects who received explanatory feedback might not have been able to apply their learning to determine the correct answer. They might have been able to employ the explanatory feedback information sufficiently to avoid the previous error. Those who in the KOR feedback group might not commit the same previous error because they had been informed "You are wrong" before. As for subjects in the KCR feedback group, because they were only provided with the correct answer, they might have easily forgotten the correct answers due to their inability to understand why the correct answer was right. This reason might have been responsible for why, with respect to the measure, probability of repeating a practice error on the delayed posttest, the explanatory feedback group scored significantly lowered than the KCR feedback group and the KOR feedback group scored slightly lower than the KCR feedback group. Again, if subjects of the study were not motivated to learn or had difficulty in comprehending the materials, whatever they read from the lesson would have been faded quickly from their memories. As a result, all the three feedback groups might have performed poorly and similarly on a post test that was delayed. This reason might have been responsible for the failure to reach a significant difference on the measures of overall correct responses and the probability of changing a practice error to a correct response on the delayed posttest among the three feedback groups. Also, it might be that because the participating subjects had failure rather than success in understanding the learning materials, they might have had less positive attitudes toward the instruction in spite of the different treatments they received. On the measure of subjects' attitude toward the instruction, the mean difference among the three feedback groups was not statistically significant, however, in terms of mean differences, subjects in the explanatory feedback group scored higher on the attitude questionnaire than did students in the KCR and KOR feedback groups, and subjects in the KCR feedback group scored higher on the attitude questionnaire than did those who in the KOR feedback group.

According to Lee (1989), the above explanations could be supported by the following evidences. The subjects involved in her study were students enrolled in the Computer Literacy course. Most students took the Computer Literacy course because it was required by their departments. Consequently, many important characteristics such as need for achievement, level of interest, attention and motivation might be different in other populations. Besides, results obtained from the pretest revealed that one third of participating subjects were considered as low prior knowledge students. It may be that because most participants did not have any prior knowledge on programming languages, their learning motivation might have been seriously suppressed. They might have even been anxious about learning this content, further depressing their motivation and performance. Comments made on the attitude questionnaire confirmed the above assumption. These comments indicated that many participating subjects felt that both the instructional materials and the practice questions were fairly hard. They had difficulty in learning from the instructional materials and seemed to have trouble with completing the practice questions during the class time. Through analyzing each test item contained in the immediate posttest during pilot study, it was found that there were two questions which were missed by every participating subject. In addition, the reliability analysis of this instrument revealed a fairly low reliability coefficient (0.6047). Therefore, the test items might have been too hard for most participants. Also, there might be some inadequacy in the test item, which was consistently missed by all participants.

For the second variable, an extra correction opportunity, Lee (1989) examined how an extra error correction opportunity (a second try) as opposed to a single try served to correct a practice error, promote posttest performance, and encourage a positive attitude toward the instruction. It was predicted that on both the immediate and delayed post tests, subjects receiving a second try upon a missed question would perform significantly better than those who received only one try upon the same question on overall correct responses and the conditional probability of changing a practice error to a correct response on a test. Unexpectedly, with respect to (a) the overall correct responses, (b) all conditional probability measures on the two posttests and (c) the participating subjects' attitude toward the instruction, the results did not support that the provision of a second try upon a missed practice would result in better performance than the provision of a single try. As a matter of fact, only the means of the attitude toward the instruction were in the predicted direction.

According to Lee (1989), one of the possible reasons for these unexpected results might be that, as discussed, the participating subjects might have been unmotivated to maintain the level of attention expected from them throughout the entire learning process. In such a case, most subjects might have passively gone through the whole lesson without paying attention to the learning content. When they were asked to do the practice questions, they might have experienced frustration due to their inability to understand the questions. They might have expected to finish the practice questions as soon as possible. If so, during the first try, most subjects might have no idea regarding how to find the correct answer. After an error, feedback might have been insufficiently remedial to correct misconceptions of rules and concepts. Thus, during the second try, even with the help of feedback, they might not have been able to eliminate the previous error or to determine the correct answer. Very possibly, most of them might have been seriously confused by the feedback information and then might have made an extra error during the second try. In this case, those who received a second try after an error might have resulted in worse posttest performance than those who received a single try following an error. Those who received a single try after an error did not have an additional opportunity to make an extra error and therefore could not repeat a previous error on a later performance. According to Lee, these findings were consistent with the findings obtained in Hodes' (1984) study in which an extra corrective opportunity upon the missed item was not superior to a single try upon the same item in raising the posttest scores.

As for the interaction effects, Lee (1989) predicted that on both the immediate and delayed posttests, the group with explanatory feedback paired with a second try upon a missed question would perform best in (a) producing overall correct responses, (b) raising the conditional probability of changing a practice error to a correct response on a test and (c) reducing the conditional probability of repeating a practice error on a test. In addition, the subjects who had receiving KCR feedback with a second try after an error would perform significantly better than those who received KOR feedback paired with a second try after an error with respect to these same dependent variables. These predictions were based upon the following expectations. First, compared to KOR or KCR feedback, explanatory feedback given after an error would better help the learner to reason out why an error occurred because it contained corrective information. After an error, if the learner was able to reason out why an error occurred and was given an extra try, he or she would process the feedback information to a deeper degree and then be able to determine a correct answer during the second try. This process would result in better learning, thereby producing more correct responses on a later performance. If explanatory feedback was not paired with a second try, the learner would have no opportunity to rehearse the feedback information during practice and might forget it easily. Secondly, compared to KOR feedback, KCR feedback given after an error would better help the learner to realize his or her mistake because it would allow him/her to compare the difference between his/her incorrect response and the correct answer. After an error, if the learner discovered his/her misconception and was given a second try, he/she might use the second try opportunity to figure out the correct answer during the second try. Such a process would prevent the learner form repeating a previous practice error on a later performance, thereby potentially resulting in a better posttest performance. If there was no second try paired with KCR feedback, the learner would have no opportunity to apply the corrected concept or rule during practice and might repeat the same error in the future. Again, the KOR feedback simply informed the learner if his/her response was correct or not. Thus, after an error, those who received KOR feedback might never know why they made such a mistake. However, they might be able to hypothesize why their errors happened. In such a case, if a second try was paired with KOR feedback, it might help those learners to discover the correct answer during the second try. Such a process may subsequently produce more correct responses on a later test. If there was no second try paired with KOR feedback, those who could hypothesize why their errors occurred would have no opportunity to try out their hypothesis during practice and might commit the same errors again. However, the results failed to support any of these predictions. One of the possible reasons for the results might be a result of the unexpected findings obtained from the analysis of the second try factor. Without the predicted effects of the tries factor, the interaction would not be expected.

In summary, from all the empirical findings and theoretical suggestions presented in this section, it might be concluded that when the task is at the higher cognitive level, delayed corrective or explanatory feedback seems to facilitate learning better. When the task is at the lower cognitive level, immediate KOR feedback appears to be more effective for learning (Carter, 1984; Gilman, 1969; More, 1969;

Sturges, 1969). And, end-of-session feedback which incorporates evaluative or corrective information seems to be more beneficial for high-mastery students. On the contrary, immediate KOR feedback appears to be more beneficial for low mastery students (Cohen, 1985; Gaynor,1981; Rankon & Trepper,1978).

More Recent Research Continuation on Feedback in Technology-Based Learning Environments

Clearly and again, from all the empirical findings and theoretical suggestions presented above, more studies were needed to determine how the difficulty of instruction, the level of effort and motivation and learners' characteristics influence the effectiveness of different types of feedback and their pairing with one or more error correction opportunities on error correction. Many more experiments and extensive literature review have been conducted to meet this need. For example, Clariana and Lee (2001) continued to conduct experiments to explore whether or not recognition-based feedback was as effective as immediate feedback and revealed that recognition-based feedback with an overt response worked better for graduate level students to learn concepts of instructional system design. Mason and Bruning (2001) synthesized the feedback literature trying to figure out the very complex relationship of feedback and other critical variables.

In addition, Mason and Bruning (2001) and Cyboran (1995) all agreed that the need and continuation of integrating feedback into a computer-based learning environment is evident. However, these researchers also pointed out that research results on the effects or the most effective type of feedback have been mixed, inclusive or sometimes, conflicting. Like most of the studies presented above in this chapter and the entire book, Mason and Bruning concluded that instructional or learning factors including the elaborative nature of feedback information, learners' level of understanding and achievement, learners' learning attitude, learning control, response certitude and timing all could have certain influence on the level of feedback effectiveness. Mason and Bruning cautioned that there is no research evidence supporting a definite type of feedback information that is the most superior one delivered by a computer-based lesson. Further, they joined many other researchers and emphasized that it is imperative to consider factors such as level of learner achievement and prior knowledge and the domain of the learning tasks involved before determining the "best' type of feedback for the learner.

In addition, Mason and Bruning (2001) reviewed studies on learners' achievement levels, depth of knowledge, attitude toward feedback, learner control and response certitude. They detailed the results by reporting the followings. Regarding how feedback effects influenced by achievement levels of learners, studies of Gaynor (1981), Roper (1977) and Clariana (1992) were reviewed. These studies indicated that there were differences among lower and higher ability learners regarding how effectively they used feedback information. Lower ability learners tended to benefit from more immediate, specific feedback such as knowledge of the correct response feedback. Higher ability learners appeared to learn more from feedback by actively processing the feedback information provided.

Similarly, from reviewing the study of Clariana (1992), it was pointed out that for learning tasks that require a lower level of understanding, delayed and knowledge of correct response feedback worked better than answer-until-correct or no feedback. For learning tasks that require a higher level of understanding, answer-until-correct feedback may have "pushed" the learner to engage in a deeper processing process for the content and therefore achieved better learning. With respect to learners' attitude toward or control over feedback, studies of Pridemore and Klein (1991, 1995), Waddick (1994) and Schimmel (1988) were reviewed. The results of these studies showed that generally, learners' attitude toward feedback or control over which type of feedback they would choose did not affect the outcomes of their learning. However, those who received little or no feedback preferred to have more feedback information while those provided with elaborative feedback did not indicate such a preference. These two researchers cited Schimmel's suggestion stating that learners with a higher level of ability are permitted to have control over their choice of certain type of feedback. This suggestion was due to that those who are more capable tend to use the feedback of their choice to produce more self-directed learning. Learners who do not have a higher level of ability may simply choose the correct response feedback and do not take advantage of feedback information to guide their own learning.

Finally, regarding response certitude, several studies were reviewed by Mason and Bruning (2001). Response certitude refers to learners' level of confidence in the selected response to a question, which include several patterns: learners may have produced a high certitude, correct response, a high certitude but incorrect response, a low certitude but correct response and a low certitude and incorrect response. For each of these patterns, the research results showed learners do not know how to benefit from feedback when they had low certitude toward a response, despite the response was correct or not. Learners may spend very little time studying feedback information when they had high certitude and a correct response (Kulhavy, 1977; Kulhavy, Yekovich, & Dyer, 1976). For learners who had a high certitude but faced an incorrect response, feedback tended to be most beneficial. Those learners in this case were motivated to find why an incorrect response was erroneous as well as to search for the correct answer.

Most of the discussions and conclusions above are drawn from feedback as a variable in stand-alone instructional lesson: print or computer-assisted or computer-based. More recently, studies have been conducted to investigate how feedback functions in a system with multiple types of technologies (multimedia) and network type of technology-based learning environment such as the Internet or the World Wide Web, WWW. For instance, Champoux's (1991) paper titled, "Designing Feedback Mechanisms into Systems to Enhance User Performance", focused on the tangible benefits of feedback. Champoux emphasized that since the 1950's, feedback's role in human motivation has enjoyed a rich intellectual history. Feedback's three major functions are, according to Champoux, to remove uncertainty, to provide sources of cues, and to provide motivation. A well-designed feedback system does all of these. Based on these functions, clearly, Champoux has employed the term "feedback" to include informative, explanatory and motivational type of information in feedback. Champoux then designated an 8-point system specially designed for creating feedback systems that covers these three major feedback functions to inform, to remove or clear uncertainty and to motivate.

In addition, Yoon, Ho, and Hedburg (2005) conducted studies showing how teachers used a combination of informative or explanatory feedback and technology to aid in guiding their students. Six teachers at a secondary and primary school used various technologies to interact with students. The teachers' names were Andrew, Belinda, Colin, Denise, Ellen, and Fauziah. They teach various subjects such as English, Math, Science, Geography, and Social Studies. The different types of software used were varied. For instance, Ellen, the Geography teacher, utilized GIS, a geographic information system, to aid her classes. Yoon, Ho, and Hedburg closely observed the teachers within each given day. Copious notes were taken, and observations were numerous. The six teachers were easily able to utilize technology with ease, proving that teachers have a high level of technology potency. In addition to the technology skill used, informative or explanatory feedback was used to help facilitate learning and conclusions. In several scenarios mentioned in the study, teachers used extensive facilitation and feedback abilities to guide students to the right conclusions. In several scenarios, the teachers showed a high proficiency in using common software, such as Microsoft Word, to help their students learn extensively about relatively complex concepts. For instance, Andrew, the language arts teacher, used Word to illustrate to students the process terms used for stories. By using simple color-coding, Andrew was able to detect mistakes in these terms and incorrect classifications of these terms. Based on the data collected, Yoon, Ho, and Hedburg concluded that it is definitely possible to meld informative or explanatory feedback and technology together in a classroom environment.

In addition, Roussou, Oliver, and Slater, (2006) also used extensive 3-D equipment and software to try to demonstrate the feasibility of the use of 3-D imaging and virtual reality in aiding children in learning. Roussou, Oliver, and Slater tried to explore the benefits of using virtual reality with full feedback systems with informative information. The participating subjects were 8 through 12-year-olds. An immersive stereoscopic VR system using a joystick was the basic technology used. The game used to teach the children about fractions involved a playground. The playground had to have the dimensions of some parts (swings, monkey bars, etc.) reduced and increased via fractions.

Questions regarding fractions had to be answered. A typical question in the simulation was, "The swings' area needs to be increased by the larger of these two fractions: 1/3 or 1/4. Which one should the swings' area be increased by?" The program also exploited common errors children made. Rather than merely showing the correct answer and giving feedback that simply informs an error, the program instead forced the child to find out his or her mistake on his or her own. For instance, the common mistake that 1/3, for instance, is smaller than ¼ since the denominator is smaller, is exploited by the simulator. The children were divided into three groups. One group used the fully immersive technology with the interactive element and informative feedback. Another group watched a "robot" perform the same task assigned on the screen. The last group used LEGO blocks to simulate the activity performed in the 3-D immersive environment. The results showed that the 3-D immersive environment with explanatory feedback was best in teaching students about fractions.

Finally and most recently, in a study by Lee, Shen and Lee (2008), a technology-assisted learning system was proved to be more effective than lectures alone to improve second graders' Chinese learning. This system was implemented based on a thorough analysis of specific errors made by the second graders in Taiwan and the learning needs that caused these errors. The system was developed to employ vivid, relevant images, graphics, animations and sounds to present Chinese words, which involves complicated pictograms, logical aggregates and/or pictophonetics for young learners in Taiwan to acquire. In the system, assistive digital boards and pens were also used to help the second graders to practice accurate Chinese stroking and writing. Moreover, corrective or explanatory feedback was provided in a way that not only indicates mistakes, but also provides vivid, correct Chinese stroking and ·

writing procedures for self-correction. The results revealed that the system, compared to the lectures, provided a much better results in helping second graders in learning phonetic symbols, performing radicals recall, and writing compacted Chinese words in an immediate posttest. The system was also more effective than lectures in helping the students sustain their good learning results in remembering meanings and writing complex Chinese words in a delayed posttest. In addition, almost all of the students who used the system indicated that they liked the system and will use the system to learn their own language again. The researchers of this study concluded that a well-designed instruction that incorporates multi-form of technologies with explanatory feedback is vital for Chinese language learning. These researchers further suggested that such a system could become beneficial in helping learners acquire similar types of subjects that include both textual and graphical forms of learning.

Summary

This chapter focuses on the many aspects of feedback and its effects. Feedback is stated to be message or communication showing either positive reinforcement or corrective messaging. Feedback, especially since the 1950s, has possessed a large intellectual following. This chapter covers the theoretical perspective of feedback, together with functions, and assumptions of feedback. Research findings concerning how feedback works with many other variables are presented. In addition, more recent research continuation on feedback is also included.

The first section of this chapter presents the many definitions and types of feedback. Then, this section covers that in the past, feedback is expected to act as a connector between responses and preceding stimuli. Most behavioral psychologists maintained that feedback was merely a positive stimuli or reward for performing a task or answering a question correctly. Skinner (1951, 1957, 1968) pointed out that if positive feedback was given for a response, the action which caused the positive response would occur again in animals. For humans, Skinner (1968) stated that immediate feedback helped reinforce desirable traits. As early as 1956, many disagreed with that theory regarding feedback's function. Ammons (1956) stated that rather than merely reinforcing good behaviors, feedback also serves an important role in correcting negative or erroneous behaviors. Cybernetic psychologists agree. In the 1960s, it was revealed that feedback was not only meant to prove or disprove an overt response; feedback is always corrective. Learning can improve if feedback is given. It is also believed that the major effects of feedback are truly internal, although they might require an external check. Later, it was pointed out that feedback fixes learning in two ways. Many others have stated that feedback has much less of a cognitive impact on positive actions than negative actions. From a constructivist learning viewpoint, feedback greatly aids in making students acutely aware of their learning environment and the content learned.

Finally, current studies have increasingly included technology as a vital component of experiments. With the invention of the microcomputer, computers were soon featured. Now, with fully immersive 3-D environments, massive storage capacities, and global internet service, educational technology has reached new heights. The internet, 3-D technologies, and new educational software all feature in more recent feedback research. Current studies focus more on uses of the internet for better feedback, consumer feedback systems, software feedback systems, and newer 3-D technology in feedback. The

increasing prevalence of intranets also motivated studies. The results from the feedback study proved that a single, specific model can be applicable to numerous feedback systems and situations. For example, 3-D technology and its useful feedback applications were explored in a study detailing how chat room technology could be the future of feedback. Results from that study ascertained that it is very possible to utilize 3-D regions online to teach and instruct students. 3-D technology has also been explored extensively by studies for the educational market. In conclusion, feedback will continue to serve an important role in instruction and learning, with more advanced technology.

Chapter Four
第四章

Feedback Processing Procedures and Suggestions
回饋訊息的了解、傳遞與建議

本章要點

　　一如前述，本書的目的在於詳介回饋訊息的諸多細節及其使用於科技教學與學習環境裡的效應。本章申論專家學者們討論學習者如何吸收、了解、與傳遞回饋訊息的諸多假設與論述。這些論述分別衍自認知學派及建構立主義的多項理論。本章並提供建議以利思索如何有效地於科技教學與學習環境裡提供回饋訊息。同理，這些建議亦源發於專家學者們從認知及建構主義思索而來的結語。

Introduction

The main purpose of this book is to discuss the functions, mechanism and effects of feedback in technology-based learning environments. The first two chapters detail what a technology-based learning environment is, the factors to consider and suggestions for designing an effective one. Chapter three covers details of feedback including the definitions, types, theoretical bases and research findings. This chapter serves to discuss, in detail, the procedure that a learner may use to process feedback. This chapter continues to provide suggestions regarding how to best integrate feedback into a technology-based lesson for increasing learning results. Accordingly, related literature on procedures proposed by researchers regarding how learners process feedback will be discussed first. Then, suggestions concerning how feedback should be provided for the best learning results will be covered.

Feedback Processing Procedures

Kulhavy's (1977) Model

Feedback processing procedures refers to the process or the procedure that a learner adopts to recognize, remember, or rehearse all the information contained in feedback in order to maintain a correct answer or correct a mistake. As early as 1977, a detailed feedback processing procedure was described by Kulhavy. This process begins with the learner processes the learning content first followed by a question. The learner will be sent down one of two paths depending upon feedback received that indicted whether or not the answer was correct or incorrect. If the answer was correct, the response confidence in the answer will be determined. With feedback indicating a correct answer, high confidence in the answer will be produced, then the learner moves on to the next frame. When the feedback indicated a mistake, low confidence in the answer will follow. Then, the learner will first study the content over in order to achieve a better understanding and then move onto the next learning unit or frame in a computer-based lesson. Given the case that the answer is incorrect and the learner's confidence level is high, then the learner will be forced to rescan the text to find the source of the error. Then, after studying that error in order to correct it, the learner will move onto the next frame. If the confidence in the answer was low, the learner will be forced to study the items to correct errors, and then be allowed to move onto the next frame. Kulhavy's feedback processing model takes into account learner confidence, which can be a major indicator of whether or not a correct response was actual knowledge. Therefore, feedback in its facilitative form and its confidence-reinforcing form plays an important role for learning. Kulhavy further emphasized that any learner could plug in content from a quick scan of material, however, in the process, the learner does not place the information into long-term memory. This will cause the learner to not truly learn the content, but instead to skim the material, answer the questions, and have a feigned understanding of the actual content. Therefore, gauging learner confidence is essential to feedback studies. Results can be altered if feedback confidence is not known or is not established. If there is no confidence in a correct response, it is most likely that the learner is guessing, and, by luck, managed to arrive at the correct answer. Therefore, response confidence is extremely important and should be included as a part of feedback studies.

Lee's (1989) Hypotheses

Lee (1989) who agreed with Kulhavy (1977) and other cognitive theorists, including Gagne (1970, 1977), Phye (1979) and Sternberg (1983), emphasized that feedback is not used solely for reinforcing a correct answer to a question. Instead, feedback is assumed to be corrective and given in order to facilitate a learner's internal processing of learning as well as aid him or her in correcting misconceptions. Lee maintained that a learner when engaging himself/herself in producing a particular response requires internal, cognitive processing of learning content as well as feedback presented. Lee, based on a cognitive learning point of view proposed by Gagne (1977) and Kulhavy's model on feedback processing procedures, extended her hypotheses. Lee explained how a learner processes learning content and any type of feedback received. First of all, Lee reemphasized the information processing process proposed by Gagne before explaining her view. Like Gagne, she explained, in general, stimulation from the environment activates a learner's receptors to produce sensory information. Such information remains in the sensory memory for only a fraction of a second. If the momentary sensory information is attended to, it enters the short-term memory, which has a very limited capacity. Information that has been deeply processed goes into the long-term memory. In other words, information that has been organized, synthesized, analyzed, rehearsed or attended to can be stored in the long-term memory and retained there for long periods of time. When needed, information stored in the long-term memory can be returned to the short-term (working) memory if it has been successfully retrieved. Accordingly, information that is meaningful is easier to retrieve and encode. As long as the retrieved information becomes readily accessible, it can activate the response generator to generate a response. Such a process then activates the effectors to exhibit the performance. As for the expectancies and the executive control, the former serve to motivate the learner to reach the goal of learning and the latter is employed to coordinate the entire flow of information processing.

Then, Lee (1989) used the following example to elaborate her view regarding how a learner processes feedback in general. Suppose a learner is practicing a rule learning question such as "Is '7$' a numeric variable?" by using a computer-based or assisted lesson. During responding, the learner would have to use his/her receptors like hearing, seeing, touching, etc. (in this case seeing) to register inputs from the outside world to his/her sensory memory. He/she would not be able to attend to all the inputs but only to those which relevant to the task. Consequently, he/she may concentrate on reading the question shown in the screen. By doing so, this question then transfers from his/her sensory memory to short-term (working) memory. In order to answer this question, he/she has to employ this question as a searching cue for activating the retrieval of accurate concepts and rules. Of course, those to-be-retrieved concepts and rules are previously learned by him/her, and are, therefore, stored and retained in the learner's long-term memory. Once the relevant rules are successfully retrieved, the learner may apply a specific rule to answer the question. While generating a response, he/she uses his/her fingers (effectors) to type in the correct answer. At the same time, the expectancies and executive control assist him/her to select the nature as well as the outputs of each processing stage in order to attend to, encode and retrieve the information most relevant to the task. Such a process directs him/her toward accomplishing the pre-determined goal of learning.

Furthermore, Lee (1989) who extended upon Gagne's (1977) belief, joined many other researchers and believed that when learners employ feedback, they usually need or require an external check for their performance. Even with the external check, learners need feedback to serve its internal effects in order to confirm, fix learning, and to make learning permanently accurate. Consequently, the whole information-processing procedure can be reexamined in light of the feedback information to make it as effective as possible. Thus, a learner could process feedback in order to confirm that learning has achieved its purpose through a confirmed correct response. In the case, feedback helps to strengthen the association between the practicing task and the related knowledge structure in long-term memory. In contrast, when a mistake has been informed by feedback to a learner, feedback could be used to provide the learner with meaningful information for the modification of the existing information-processing process. In other words, feedback could be employed to inform the learner as to whether he/she has attended to, encoded, and retrieved the information accurately. Feedback can further direct him/her to revise every stage of information as well as the entire flow of information processing

The Processing of KOR Feedback. Specifically, Lee (1989) described the processing procedures for each of the feedback type. In her study, KOR feedback only includes messages such as "You are right" or "You are wrong" to inform the learner whether his or her response is right or wrong. To process this type of feedback, a learner first reads a practice or a test question, again, which may appear like the following:

Which of the following is the name of a numeric variable?

1. C$

2. HI

3. "7"

4. "BICC"

5. 34

Your Answer (1, 2, 3, 4, 5) is

Press the return key to continue

Facing the question above, the learner began to retrieve appropriate concepts, recall appropriate rules and determine specific rules to apply. If the learner is able to go through these stages of processing, he/she then answers the question. If not, the learner may guess at the answers during any stage of processing. After responding to the question, the learner receives nothing but a simple message stating either "You are right" or "You are wrong". Accordingly, in order to answer the question illustrated above, first, the learner has to retrieve the concept of variables. He or she then has to recall the differences between the string variable and the numeric variable. Once he/she has recalled the concept of a numeric variable, he/she must recall the rules for naming variables. Then, he/she has to remember the syntax rules for writing a numeric variable. Whenever the learner is able to apply such a correct rule for naming a numeric variable, he/she would choose number 2 as the correct answer. The KOR feedback does not include any information to explain why a response is right or wrong. Neither does it indicate what the correct answer is. Hence, when an error occurs and the learner is willing to correct it, he/she has to hypothesize why his/her response is wrong. Then he/she himself/herself has to discover what the correct answer is, and to reason out why the correct answer is accurate. For example, suppose after responding to the question on a numeric variable, the learner is informed about an error and is willing to correct it. He/she has to hypothesize why his/her error happened and to figure out the correct rule for naming a numeric variable. Unfortunately, if an error is due to a lack of information on numeric variables or an unlucky guess, the learner may not be able to determine why number 2 is the correct answer.

The Processing of KCR Feedback. Next, Lee (1989) described that knowledge of correct result (KCR) feedback provides what the correct answer is, which means that KCR feedback conveys the correct answer to the learner no matter the response to a question was right or wrong. The Learner realized whether or not his/her response to a question was correct by comparing the response with the KCR feedback provided. In the case of processing KCR feedback by facing the same question illustrated above, the learner will have to go through the initial stages of processing that are identical with the processing procedures with the KOR feedback. Yet, with KCR feedback, after responding to the question, the learner receives the feedback stating the specific correct answer. In the case of an error, if this error is not due to an unlucky guess, KCR feedback may provide basis for self-correction due to the fact that the learner can use KCR feedback as a clue to infer his/her misconception. For example, facing the same question, the learner determines the fifth choice as the correct answer. Before giving the response, the learner may already recall that all variable names must start with a letter. As soon as he/she knows "Number 2 is the correct answer", he/she may try to compare the similarities as well as contrast the differences between the second and the fourth choices. He/she may, therefore, recall or become aware of the fact that no variable name is enclosed in quotes. As a result, he/she became to understand why his/her response was incorrect and reasoned out why the correct answer was correct. However, when a mistake is a result of a lack of knowledge or breakdowns in the processing of information on variable, the learner himself/herself is unable to discover why he/she made an error. In such a situation, the error may occur again in the future.

The Processing of Explanatory Feedback. Lee (1989) explained that in her study, explanatory feedback was designed to include KOR feedback to inform an error in combination with more informative messages which explain why an incorrect response is erroneous. In her study, explanatory feedback did not contain the correct answer; it informed an error and explained why. In other words, Lee stated that the explanatory feedback was primarily for providing the meaningful information that allowed learners to correct themselves or functioned as a form of remedial instruction. An example of an explanatory feedback for the same question illustrated above is shown below.

> You are wrong.
> 34 is just a numeric, so it is a numeric constant or value. All variables name start with a letter but string variables end with
> a $, whereas numeric variables do not. No variable name is enclosed is quotes.
>
> Press the return key to continue

Lee (1989) explained that after an error, the learner received the explanatory feedback that included an explanation for his/her incorrect response. When such an error is due to the learner's breakdowns in the processing of numeric variables or variables, explanatory feedback given to him/her may increase the ease with which the learner may retrieve the relevant rules for naming a numeric variable. Even when an error is due to the learner's misconceptions, he/she can rectify his misconceptions through studying the instruction supplied by the explanatory feedback. For example, when the learner erroneously recalls a numeric variable has to end with a $, he/she can correct such an error by studying the explanatory feedback which states a numeric variable does not end with a $. Consequently, the explanatory feedback makes the learner truly realize why an error is produced; therefore, the learner can be prevented from making the same mistake in the future.

KOR, KCR and Explanatory Feedback with an Extra Corrective Opportunity. Lee (1989) further described the processing procedure when a learner receives different types of feedback plus an extra corrective error correction opportunity. Learners who miss a practice question, receive KOR feedback and was asked to try the missed question again may first hypothesize why such an error occurred. During the second try, the learner may test if his/her hypothesis is correct or not. This process may subsequently produce the correct answer during the second try.

Graphic 9 : How a Learner Processes KOR Feedback with an Extra Correction Opportunity

Very possibly, when KCR feedback and an extra correction opportunity are offered after an error, the learner may just copy the correct answer from KCR feedback during the second try. If this is the case, KCR and a second try may not produce any effect on error correction. However, if the learner is willing to think over why he or she made an error, he/she can use KCR feedback to detect the error. He/she can compare the differences between his/her response and the correct answer and find out why he/she made a mistake. Once the learner himself/herself is able to infer his/her misconceptions and to rectify them, the second try can serve as a review opportunity in which the learner may attend to the correct answer to a deeper degree. This process may promote learning. Unfortunately, when the learner is unable to infer or correct the error, a second try may become redundant. When explanatory

Graphic 10: How a Learner Processes KCR Feedback with an Extra Correction Opportunity

feedback and a second try opportunity are provided after an error, learners are allowed to learn from the corrective information in the feedback during the second try.

Once the learner answers the missed question correctly through the second try, he/she can confirm that he/she has rectified his/her misconceptions of the learning materials. In contrast, when a second try opportunity is not available, learners find no opportunity to tryout the information acquired during the previous explanatory feedback. Without practicing the correct rule through another try, learners may forget this rule within a very short period of time. If this is the case, explanatory feedback may turn out to be ineffective in correcting previous errors. On the other hand, if the explanatory feedback is not provided, learners may be unable to discard their false hypotheses while responding to

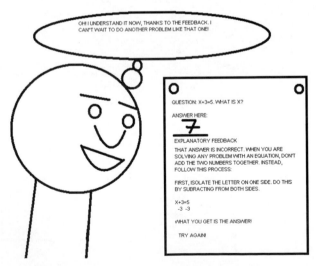

Graphic 11 : How a Learner Processes Explanatory Feedback with an Extra Correction Opportunity

the missed question again. In this case, even though the missed item is reviewed according to an increasing ratio, the learner might never discover the correct answer. Thus, the error may repeat itself in the future.

According to Lee (1989), the above arguments can find their strong support among most theoretical positions indicated earlier in this chapter. For example, Ammons (1956) asserted that only if feedback informs learners about the amount and direction of their errors, can it produce effects on error correction. Evidently, explanatory feedback can fully explain why an incorrect response is erroneous. And, in order to boost the effect of explanatory feedback, an adjoining corrective opportunity in which learners can use the explanatory feedback to search for the correct answer is necessary. In addition, explanatory feedback with a second try can perform similar functions as does a feedback-control system, which has been described by cybernetic psychologists (Smith & Smith, 1968). In cybernetic theory, a feedback-control system is crucial because it can dynamically function to generate actions toward a target, to detect errors and to use the error signal to modify the whole system. Similarly, explanatory feedback can allow the learner to compare the learning result with his/her learning result so as to detect errors. When the error is located, during second try the learner may try to eliminate his/her error and thereby modify the whole learning process.

Again, Lee (1989) cited Craik and Lockhart's (1972) levels of processing theory and maintained that it would be easier for a learner to retain information when it is processed in a deeper manner. Information in sensory storage, if not processed, leaves only a momentary sensory impression. Information that is encoded can be held in long-term memory. Long-term memory cannot be easily disrupted and better retention occurs when information is processed to a greater degree. As a result, effective instructional strategies for teaching retrieval often emphasize meaningfulness, rehearsal and review. With this view, it thus argued that explanatory feedback which highlights critical aspects that may have been misunderstood can be more meaningfully processed. When explanatory feedback is given after an error, learners are allowed to replace information that may have been miscoded with the accurate information. At the same time, if a second try is available, learners are provided with additional opportunity to rehearse the accurate information as well as to put the information in action, thereby enhancing retention.

Suggestions

Many researchers offer suggestions with regards to how to provide effective feedback in a technology-based learning environment. The followings are a number of researchers who provided detailed suggestions for this purpose.

Lee's (1989) Suggestions

Lee (1989), based on her view and the findings, which were detailed in the previous and this chapter of this book, provided the following suggestions:

- First of all, as always, learners' level of learning motivation, ability and need for achievement should be seriously taken into consideration when investigating the effects of different types of feedback.
- Secondly, the major function of explanatory feedback was to direct the learner to locate an error and, further, to provide cues for searching the correct answer. However, if there were too much information in the explanatory feedback, it might have been

too detailed to be assimilated into the learner's ongoing information processing. If this were the case, explanatory feedback might have confused the learner and might not have been as corrective for him/her as the researcher intended. Therefore, future research is necessary to determine the sufficient and appropriate amount of feedback information that can maximize learning.

According to Lee (1989), instructional designers are recommended to consider reconstructing informative, corrective or explanatory feedback into two specific types: (a) the corrective feedback, which contains corrective information and is primarily for helping the learner to locate an error; and (b) the remedial feedback, which contains brief instruction as to hint the learner to search for the correct answer. It is also recommended that both the corrective and remedial feedback should be included in any technology-based lesson, and learners should be allowed to choose the specific type of feedback they really need. Accordingly, the comparative benefits of these variations of feedback on error correction should be examined. Additionally, how each variation of feedback helps increasing the probabilities of (a) correcting an initial error to a correct response on a later performance and (b) maintaining an initial correct response on a later performance, and (c) decreasing the probability of repeating an initial error on a later performance should be determined.

In addition, for the factor of providing an extra correction opportunity, it was expected that after an error, the extra opportunity could be employed as both a corrective and a review opportunity in which the learner could process the learning materials or the feedback information to a better degree, therefore facilitating learning. However, Lee (1989) cautioned that based upon the findings of her study, a second try might have become unnecessary when a learner was unable to

understand the practice questions. Thus, Lee suggested that the influence of the level of difficulty and complexity of the instruction and practice questions on the effectiveness of a second try opportunity should be examined regardless of what type of feedback was provided. The designer of a technology-based lesson also needs to research into the question of whether the instructional event such as a second try provided during learning or practicing is necessary.

Similarly, Lee (1989) further suggested some important conditional probability measures should be considered. The suggested conditional probability measures may include the probabilities of (a) maintaining a correct response made during a second try on a later performance and (b) repeating an error made during a second try on a later performance. These measures can provide information as to whether a second try can help learners to change a previous error to a correct response and to maintain that correct response on a later performance. Finally, Lee's (1989) emphasized that in order to make the combination of explanatory feedback with a second try work, it would be necessary to determine how much and what kind of corrective information paired with a second try can really make a difference compared to simple KOR or KCR feedback on error correction. Future research is also needed to investigate when and why feedback paired with a second try should be given when considering other important variables such as nature, type, difficulty level of learning tasks, and learners' prior knowledge, characteristics and motivation.

Mason and Bruning's (2001) Framework

Mason and Bruning (2001) investigated the different types of feedback and variables necessary for consideration. According to them, feedback is a significant factor in motivating further learning, and that feedback's main functions are to help facilitate awareness of misconceptions and errors. However, they did caution the fact that feedback provided in a computer-based or technology- based system or lesson is limited by the competence and the ingenuity of the instructional designers who implemented feedback in the system.

Thus, based on Mason and Bruning's (2001) framework, the achievement level of the learners and the nature, type or domain of the learning content are the two most crucial variables to consider when providing different types of feedback. For a weaker-ability learner, immediate feedback pointing out mistakes was recommended. For higher-level learners, feedback that is delayed could be more beneficial. The reason, learners who hold a higher level of achievement are often at a stage in which they can try multiple times for a result and most probably get the correct answer within three tries. Thus, feedback that is provided after these tries could have more facilitating effects. It is also important, that feedback be tailored to the content being learned. For instance, if the computer-based lesson is about content acquisition, immediate feedback could be more effective, since it can remediate errors generated by the user. For more complex learning domain or tasks such as comprehension and abstract reasoning, delayed feedback is suggested since it allows the learner to explore many different possibilities before facing the feedback

Mory's (1996) Suggestions

According to Mory (1996), feedback functions evolved from Skinner's views that feedback was merely for positive reinforcement to the view that feedback is for correction and verification of answer purposes. She pointed out that almost all the early studies conducted regarding feedback fit into the basic overall philosophy of objectivism. As Mory explained, objectivism believes that thought is not based upon experience, but upon the external reality. To objectivists, meaning corresponds to the categories of thought in the world, and external reality is represented by symbols. The objectivist attitude towards feedback is that feedback is supposed to correspond with external reality and is primarily to reinforce a reality that has a desirable result.

Based on objectivism, as pointed out by Mory (1996), most early feedback studies focus on what is termed outcome feedback and have ignored the idea of aiding learners in their own self-guidance. Self-regulation is the processing of information based on knowledge, experience, and beliefs of a learner. The results from self-regulation include cognitive and affective gains of a learner. Lens models are often used in regards to self-regulating learning research. A lens model is a model in which a learner progresses and content characteristics are used to predict performance. Mory stated that both cognitive validity, cognitive functional feedback are information drawn from an external source describing that source's relations to a task's hints and achievement levels. Cognitive validity feedback is information stating the learner's perspectives regarding hints and the relationship of achievement. Functional validity feedback states the relationship between the estimation of the learner's achievement and the actual achievement attained by the learner. Accordingly, the effectiveness of examining feedback from a lens-model perspective is that it helps researchers realize that the effectiveness of feedback relies on the learner characteristics. Often, a learner's beliefs and misconceptions can hinder the effectiveness of feedback. These factors should be factored into account when designing feedback information.

In addition, Mory (1996) noted that with constructivism, feedback research has been regarded differently from the usual objectivist perspective. Mory emphasized that, in contrast to objectivism, by constructivism, reality is determined by the learner. Rather than merely processing symbols, the mind acts as a constructer and a former of symbols. Meaning is not determined by reality, but by the learner who constructs reality. Rather than being representative of external reality, symbols are the tools used by the mind to construct an internal reality designed to increase understanding. By Mory's suggestions, the constructivist view on feedback is that feedback's purpose is to aid in constructing an internal reality, and facilitates knowledge construction. Feedback's main purposes are to aid learners in building symbols, place in context of human experience, and to provide construction mentally. By constructivism, rather than merely using feedback as a tool for verification and correction, feedback should be used as a sort of coaching mechanism for construction of an internal reality. In other words, constructivist feedback is supposed to help the learner construct an internal reality, help the learner solve complex contextual problems, act as an aid in social negotiation, and provide guidance and challenge.

In summary, Mory (1996) provided the following suggestions for providing feedback based on constructivism:

- Feedback should guide the learner towards his or her personal reality. In other words, it must aid the learner in constructing an internal reality by providing intellectual tools.
- Feedback must aid in building symbols that hold meaning for the learner.
- Feedback must be placed in context with real-world experiences. Feedback must also derive meaning that can be determined by internal understanding.

- Feedback should be provided in such a way that would be attained through one's peers as learners solved problems in a group, team-oriented manner.
- Feedback should have the function that could act like a coaching mechanism that scrutinizes strategies used to solve problems.
- Feedback also should help the learner solve problems within relevant settings.
- Feedback should be provided to help, suggest or guide learners to understand the content from different perspectives, to form conceptual interrelation and to build many different perspectives about the content.

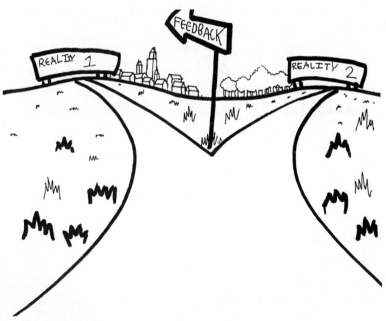

Graphic 12: Feedback Functions in Multiple Realities

- Feedback should be offered to allow learners obtain information about the object of his/her choice. A feedback that provides a continual stream of information that aids learners in establishing and maintaining goal setting and goal monitoring is recommended. Thus, feedback that can be used to judge and compare against a learner's individually designed learning goals should be offered.
- Feedback must challenge people into developing.

More Pertinent Suggestions

Consistent with Mory's (1996) suggestions, Choitz and Lee (2006) also emphasized that explanatory or informative type of feedback should be provided based on the theory of constructivism. They offered the following rationales. Constructivism is an emerging approach that provides an alternative for Instructional Designers. Most constructivists believe that learning is a process and should be embedded in realistic contexts. Therefore, learning is best manifested in open-ended experiences, and flexible approaches to learning are valued. Thus, most constructivists advocate the opinion that learning is enhanced and buffered by social interaction. Multiple representations of concepts being learned are essential because these add realism that facilitates acquisition of the concepts. Learners need to be acutely aware of their learning process and frequently encouraged to reflect on growth, goals, and needs.

Based on constructivism, Choitz and Lee (2006) pointed out that constructivism also has applications in the field of asynchronous text based computer conferencing, which refers to the use of computer software to enable learners and instructors to converse online and can be regarded as one type of technology-based learning environment. The use of this environment, according to Choitz and Lee, greatly reflects on the constructivist theory based on the following reasons. First, this environment permits a very flexible schedule allowing learners to learn by using its bulletin board format at the time and location that are convenient to them. This feature helps motivate learners to actively engage in the learning process of their choice. This feature also permits a new way of knowledge acquisition among learners: knowledge is no longer regarded as something that is given to them passively; rather, knowledge is the asset that learners can choose a convenient time and format to actively pursue it. Based on this rationale, Choitz and Lee (2006) maintained that explanatory or informative type of feedback that is provided to guide learners to understand the content, to learn from different perspectives and to construct conceptual interrelation for the learning content and to build many different perspectives about the content should be provided based on constructivism and should be offered in any technology-based learning environment, including the one featuring asynchronous text based computer conferencing.

Finally, according to Lee, Chalmers and Ely (2005), in a technology-based lesson, as mentioned in Chapter Two, learners need to be kept engaged, interested and motivated. To accomplish this, attention needs to pay to both content and interface design. In addition to the provision of effective feedback, these researchers suggested an increased level of interactivity between the learner and the events provided by the lesson, a great variety of learning events, meaningful assignments, and by permitting learners to have access to detailed, explanatory or informative type of feedback. Such type of feedback would allow learners to have an opportunity to realize their levels of success in response to the activities or events performed. This feedback would further serves the function of a remedial or review lesson in which the learners would process the provided information to a greater degree so that a self-correction process for a misconception or an learning mistake can occur (Evia, 2004; Fitch, 2004; Furnas, 1997).

Summary

The science of providing appropriate feedback is paramount nowadays in order for truly effective educational opportunities. Thus, this chapter focuses on how learners process feedback information and how to offer the most effective feedback in a technology-based learning environment.

To begin with, the researchers of the 1980's to look in on the feedback paradigm were Kulhavy (1977) and Lee (1989). Kulhavy postulated that feedback processing procedures refer to the procedures used to reinforce and foster learning via error in students. Kulhavy proposed a complex structured system for this process, with multiple stages. Then, Lee's (1989) view was reviewed. Lee stated that feedback's primary purpose is not Skinnerian; on the contrary, it is supposed to be corrective in intention and focused on facilitating understanding and true learning, not merely rote memorization. Lee's statement regarding feedback systems draws from Gagne's (1977) information processing theory that learning feedback systems are much like processing units. The processing unit takes in the information inputted by the learner and provides an answer-specific response. Lee used examples in which learners were obligated to find specific variables and identify which were independent and which were not. From these results, Lee deduced that information that is stored in the brain for long periods of time can only be used and pulled out if the learner actually understands the information presented. With the examples that Lee used, she stated that the senses of hearing, sound, and viewing all had to work in cohesion to answer some questions. Lee also emphasized that feedback could often come from external factors outside of the instructor's control, such as other learners, direct stimuli, and reactions from other factors. Learners, Lee stated, could reflect and inference an answer was correct based on cues from the outside world. Lee also described that with KOR and KCR feedback, learners can determine their mistake via merely realizing an error had occurred and checking the correct answer by the feedback provided. With explanatory feedback in which the learner is given a tailored response, based on the error he/she made, the learner receives the following benefits. Each response is given its own specific informative or explanatory feedback, which greatly streamlines the feedback process and requires no inference guess-work, which can lead to a possible misconception.

Next, this chapter moves to the many suggestions proved by a number of researchers. Lee (1989) emphasized that an extra correction opportunity paired with the explanatory feedback is necessary for a learner to engage in a self-correction process. Lee further suggested that the research question regarding how explanatory feedback in combination with an extra correction opportunity work to correct a mistake or maintain an initial correct response should be answered. After Lee's work is mentioned, then the work of Mason and Bruning (2001), in which Mason and Bruning conducted an extensive literature regarding feedback required for consideration was covered. Mason and Bruning maintain that feedback research is a potent and important field and concluded that the achievement levels need to considered and then, the informative feedback provided should be tailored to the content to be learned.

In addition, Mory's (1996) view and suggestions were presented. Mory first pointed out that many feedback studies have not taken into account the fact that aiding learners in self-guidance can also be considered as helpful. Self-regulation, defined as the processing of information based on previous knowledge, experience, and beliefs is helpful to learning. Furthermore, Mory emphasizes that, in contrast to objectivism, where the learner uses his or her preexisting realities, constructivism uses a new reality constructed by the learner. Mory stated that feedback should guide the learner towards his or her personal reality, that feedback must aid in building symbols that hold meaning for the learner, and that feedback must be placed in context with real-world experiences. In addition, Choitz and Lee (2006), and Lee, Chalmers and Ely (2005) suggested that informative or explanatory type of feedback should be provided in a way that facilitates the formation of different perspectives toward the content, the enhancement of problem solving and error correction skills.

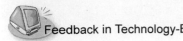

Chapter Five
第五章

Conclusion
結　語

本章要點

　　本章總結此書各章節的要點並探討如何延伸回饋訊息的未來實証路線。基此，本章重申何為回饋訊息及科技教學與學習環境。本章並回溯各章節的論點，這些論點涵蓋回饋訊息的定義、類型、理論根據、及其使用於科技教學與學習環境裡的效應。最後，本章強調設製高效應的回饋訊息的重要性並探討延申其實証效應的可行性。

Introduction

Feedback is an important component of learning. It serves to identify both a correct and incorrect response and can be employed to further educate and correct learner errors. Thus, the level or amount of information contained in feedback can range from a simple conformational message to a more extended, remedial type of mini-instruction to help correct misconceptions. With feedback, numerous opportunities for research and instructional development exist. This book contains five chapters, each written in regards to feedback definitions, mechanism, research and suggestions for the provision of effective feedback. The chapters all cover different aspects of feedback in technology-based learning environments, and include many different and functional views.

Concluding Remarks

In this book, the construction of viable and functional technology-based learning environments was first covered. The definition of technology-based environments is then provided. The terms computer-assisted instruction, computer-based instruction, and computer-managed instruction are all discussed and clarified. However, it is also emphasized that, in this era, these programs are largely not broad enough to categorize many different programs which utilize networked systems, such as the internet and private company intranets. It is stated that nowadays, computer learning systems are almost always based on networks, rather than on stand-alone systems. Next, the fact that numerous technologies are commonly used in technology is stated, and the important emphasis on the fact that technology should not drive learning or the design or the instruction. In addition, it is stressed that not only one kind of technology should be used in a technology-based learning environment. Rather, pertinent theories, paradigms, technologies, tools, equipments, materials or even resources that can be used to deliver and enhance learning are all considered to be appropriate in a technology-based learning environment.

After that, the different kinds of computer technology used in technology-based environments are specified. The input technologies such as the mouse and the keyboard are used to input information and commands. Output devices, such as printers and speakers, convert data in the computer into a form which can be accessed by others. In addition, numerous types of computer-input-output technologies such as CD-ROMs, flash drives, floppy disks, memory sticks, and numerous others are covered. With network modems, game controllers, and other input and output technologies, technology-based learning environments have a plethora of options regarding technological choice. In addition, hand-held computers, such as palm pilots, mininotebooks, and internet-capable cell-phones, such as the Iphone, all offer potential for new learning systems.

After the hardware used in technology-based learning environments is discussed, software, its meaning, and its purpose in technology-based learning environments is discussed. Computer software refers to programs which introduce content and provide various functions. Software is the element in a computer system which is the actual interface, the actual component used to make the united components come together to form the cohesive whole. Software is what directs the computers to function appropriately and make correct decisions based on the data inputted. The specific function of software in technology-based learning environments is to tutor, drill, simulate realities, and aid in the acquisition of knowledge. In addition to software-based learning systems, internet learning systems, either using one-on-one chatting rooms or massive forums, are gaining in popularity, with many being created.

Internet forums and one on one internet classrooms rely on the technology of the internet. The internet is a global information exchange system with a series of systematically linked computers which transfer information all over the world. Next, the difference between the internet and the WWW was explained. The internet was a military forum originally, in existence since the 1960s. The WWW has become the dominant entity for transferring information. The fact that standard terminology, even updated terminology designed to reflect the needs of future learners, cannot effectively be applied to internet learning stems from the fact that internet learning is so much more diverse than computer hardware. Some terms include e-learning, WBL, synchronous WBL, and asynchronous WBL. Furthermore, it is noted that interactions between the instructor, and the learner can often occur without time and location restrictions. LMS and LCMS systems are used to view learner statistics and content data.

With three different learning types to choose for training, and numerous different options for technologies to be used, it is important that the right decisions are made about such topics. Commonly, in the first method of instruction, face-to-face, micro and macro strategies, strategies focus on both the specific minutiae of content and the big picture. Next, e-learning, learning entirely within the setting of a computer network, relies on technology. Finally, blended learning is the system in which technology and face-to-face instruction is combined.

There are numerous factors that need to be considered when one needs to purchase a module, including cost, efficiency, and content consistency. Expenses for a WBT can range anywhere from 75,000 to 1.6 million dollars. Content on a WBT needs to be tailored specifically for the WBT format. Complex content is highly unsuitable for WBTs, as it often requires explaining from an on-site facilitator, something not available in e-learning. Furthermore, the content must remain consistent. Also, problems with technology, such as low bandwidth and high-intensity content such as video clips, need to be prepared for. Learners need to also be considered. Learner technological literacy is critical to the success of the lesson, and needs to be investigated before designing the WBT.

Before one even conceptualizes a finished WBT or a technology-based module, a design model must be selected. The needs analysis is necessary in that it determines what the learners need in the lesson. It shows learners' strengths and deficiencies as a whole, and demonstrates what is necessary for focus in a technology-based module. In addition, strong teams, made up of multiple components, are necessary, if not imperative for the design of a technology-based lesson. These teams should have varied talents which complement each other. Furthermore, learner analysis is always needed to determine the learners' level of technological literacy, content knowledge, and general status. Another challenge that is commonly faced by many designers is to keep the learners interested in WBT for the duration of the class. With the endless allure of surfing the web, it is often quite difficult for designers to keep learners engaged with the modules. Content analysis is often helpful in making the content more usable and more interesting. Also, chunking content appropriately is necessary if not imperative for learners.

Next, the definitions, types, and functions of feedback in technology-based learning environments are covered. Feedback, first off, is defined as the process in which the factors that produce a result are themselves modified, corrected and strengthened, by Webster's New Unabridged Dictionary. This definition of feedback as an agent for corrective change has been updated over the years to include numerous types of feedback. However, the three most commonly used kinds of feedback are knowledge of the response, knowledge of the correct answer, and explanatory feedback.

The first conception of feedback as an educational concept was that feedback served only as an agent to reinforce correct answers. Thorndike (1911, 1913), Skinner (1951, 1957, 1959, 1968), and other early behavioral psychologists all stated that feedback was designed to be a positive reinforcer. However, as early as 1956, Ammons disagreed with the perception that feedback was only for positive reinforcement. Furthermore, cybernetic psychologists disagreed with the viewpoint that feedback was only for positive reinforcement. Instead, feedback should function as a corrective agent, information designed to detect errors, find misconceptions, and most importantly, correct them. It has been stated that feedback serves as a cognitive tool based on its clarifying and correcting capabilities. Furthermore, feedback, through this, helps clarify correct answers as well as increasing confidence in later answers. Also, feedback can become anything from a simplistic yes/no answer all the way to extensive research complete with new material. With error correction, feedback steps in to impact cognition by causing learners to question their previous comprehension of the content presented.

Based on a behaviorist point of view, feedback should be provided after every correct answer. Therefore, early studies were based on this principle. Studies ranging from Buss and Buss's (1956) early foray into feedback research all the way to Rosseau, Oliver, and Slater's (2006) studies into fully-immersive Virtual Reality studies on feedback's effects on child learners all ask the same question: How does feedback work, aid, and support learners?

With the effects of feedback, there are numerous ideas provided. The primary ideas motivated by the studies of Buss and Buss (1956), Sassenrath (1975), Bardwell (1981), and Moore and Smith (1964) have relevance to this day. Their ideas revealed that feedback does not work solely for a confirmational purpose. If feedback only acts to confirm a correct response, immediate feedback would produce the best effect on learner performance. However, these researchers revealed that delayed feedback, and corrective feedback allowing more time for cognition and error recognition, have better effects and generally cause learners to perform better.

There have been debates regarding whether feedback should be administered immediately or delayed. The studies conducted by numerous researchers have conclusively proved that there is a relationship between the timing of feedback and the performance of learners. Some studies, such as those conducted by Cohen (1985) and Rankin and Trepper (1978), stated that high-level learners benefit more from delayed feedback than immediate feedback. Some more conclusions can be drawn, though. For high-mastery learners, delayed explanatory feedback, as it gives more time to process content and draw conclusions, is more effective than immediate feedback for high-mastery content. On the contrary, for low-mastery learners working on low-mastery content, immediate KOR feedback works better overall, rather than delayed feedback. In addition, research into feedback scheduling has concluded that learners who used constant KCB feedback learned better than learners with no feedback.

When feedback is combined with other variables, feedback gains a level of uncertainty. Clearly, the amount of information contributed in feedback is essential to learning. However, conflicting findings have revealed little regarding the true nature of feedback information. For instance, studies involved college students given either minimal information feedback on a low-mastery task or extensive information. The studies proved that the minimal feedback proved more so effective than the high-information feedback. However, other studies contradict this. It is apparent that feedback information provided should be investigated with additional variables such as learners' prior knowledge, the difficulty level of the learning content or learning tasks, learners' motivation level and learning characteristics.

Next, feedback processing procedures are covered. Feedback processing refers to how learners process and view feedback given to them. Feedback processing is a multi-faceted process. For example, Kulhavy (1977) created a complex, multi-faceted feedback information processing model that details how learners process feedback information under different response and response confidence situations. Based on Kulhavy's model, Lee (1989) and Mason and Bruning (2001), also suggested that feedback is a significant focus and need in learning systems, and that its primary purpose was to find errors and facilitate their correction. Lee stated that learners are similar to information processing systems, with the eyes and other senses being input devices, the brain being the CPU, and speech or writing being the output devices. Feedback is the information being inputted into the central processing unit. If the feedback is not well-constructed, and if the information is misapplied, then the information is utterly useless. Therefore, Lee stressed the importance of quality, informative or explanatory type of feedback.

The processing of quality, explanatory feedback involves a rather complicated process, including recognizing a mistake, understanding the explanation provided by feedback, and realizing how to employ the explanation obtained from feedback to engage in a self-correction process. When explanatory feedback is coupled with an extra corrective opportunities, the learner can consider the explanation and then retry the question, potentially resulting in a better learning results. The suggestions that Lee (1989) offered for feedback research included investigation into learner confidence and motivation, using feedback to aid in finding errors, and dividing feedback into corrective and remedial types. These two pairs of feedback, Lee stated, are necessary in any technology-based lesson. Furthermore, Lee stated that an extra corrective opportunity should be provided when it can be employed as both a corrective opportunity and a review opportunity. Lee also suggested that a second try needs to be gauged in order to determine whether or not it will aid in the detection of errors, the general correction of errors, and the retention of correct answers.

Furthermore, with constructivism, learners construct their own alternate learning realities. In a constructivist environment, feedback is more important than ever. With flexible approaches in learning and innovation, feedback is necessary to guide the learner towards his or her own reality, aid in building symbols for the learner, help place experiences in context with real-world events, deriving internal understanding, and serve as a mental construction tool. With this, Mory (1996) and Choitz and Lee (2006) all suggested that feedback, firstly, should guide learners towards the realities most apparent for him or her. Feedback must also aid in building symbols which hold meaning for the learner, must be placed in context with existing world events, must be presented in a group-oriented manner, must always help in a positive way, and most of all, must challenge learners into developing.

125

Finally, it has been indicated clearly by the lack of conclusive evidence that more research into feedback is necessary. Even with the technological advances of the modern age, feedback is still one of the major components of effective learning systems. With the coming of the Information Age, feedback research in new areas is essential. With the development of immersive 3-D reality systems, multimedia learning programs, and use of a virtual learning form in cyberspace, studies into feedback using the latest technology are more prevalent than ever. Therefore, feedback research is ever-expanding.

References
參考文獻

Alessi, S. M., & Trollip, S. R. (1991). *Computer-based instruction: Methods and development* (2nd ed). Englewood Cliffs, NJ: Prentice Hall.

Ammons, R. B. (1956). Effects of knowledge of performance: A survey and tentative theoretical formulation. *The Journal of General Psychology, 54, 279-299.*

Anderson, J. R., Conrad, C. G., & Corbett, A. T. (1989). Skill acquisition and the LISP Tutor. *Cognitive Science, 14*(4), 467-505.

Anderson, R. C., Kulhavy, R. W., & Andre, T. (1971). Feedback procedures in programmed instruction. *Journal of Educational Psychology, 62*, 148-156.

Ausubel, D. P. (1968). *Educational psychology: A cognitive view.* New York: Holt, Rinehart and Winston, Inc.

Bardwell, R. (1981). Feedback: How does it function? *Journal of Experimental Education, 50*, 4-9.

Beer, V. (2000). *The web-learning field handbook: using the World Wide Web to build workplace learning environments.* San Francisco, CA: Jossey-Bass.

Berners-Lee, T. (1989) The original proposal of the WWW, HTMLized. A hand conversion to HTML of the original MacWord document, retrieved 21 December 1997, from http://www.w3.org/History/1989/proposal.html.

Bloom, B. (1956). *The taxonomy of education objectives*. New York: McKay.

Bober, M. (2001). Circuits and Systems for Video Technology. *IEEE Transactions, 11*(6), 716-719.

Bonk, C. J., & Reynolds, T. H. (1997). Learner-centered web instruction for higher-order thinking, teamwork, and apprenticeship. In B. H. Kahn (Ed.), *Web-based instruction* (pp.167-178). Englewood Cliffs, NJ: Education Technology Publications.

Brown, J. S., Collins, A., & Duguid, P. (1989). Situated cognition and the culture of learning. *Educational Researcher, 18*(1), 32-42.

Buss, A. H., & Buss, E. (1956). The effect of verbal reinforcement combinations on conceptual learning. *Journal of Experimental Psychology, 52,* 283-287.

Carroll, A. B. (1976). A three-dimensional conceptual model of corporate performance. *Academy of Management Review, 4*(4), 497-505.

Carter, J. (1984). Instructional learner feedback: A literature review with implications for software development. *The Computing Teacher, 12*(2), 53-55.

Chalmers, T., & Lee, D. (2004). Web-based training in corporations: organizational considerations. *International Journal of Instructional Media, 31*(4), 345-354.

Champoux, J. E. (1991). Designing feedback mechanisms into systems to enhance user performance. *Journal of Systems Management, 42*(8), 28-30.

Chalmers, T., & Lee, D. (2004). Web-based training in corporations: organizational considerations. *International Journal of Instructional Media, 31*(4), 345-354.

Chanond, K. (1988, January). The effects of feedback, correctness of response, and response confidence on learner's retention in computer-assisted instruction. Paper presented at the annual meeting of the Association for Educational Communications and Technology, New Orleans.

Choitz, P., & Lee, D. (2006). Designing asynchronous, text-based computer conferencing in the corporate environment: Ten research-based suggestions. *Performance Improvement Quarterly*, *19*(3), 55-71.

Clariana, R. B. (1992). Integrated learning systems and standardized test improvement. Unpublished doctoral dissertation, University of Memphis, Memphis, Tennessee.

Clariana, R. B., & Lee, D. (2001). Recognition and recall study tasks with feedback. *Educational Technology Research & Development*, *49*(3), 23- 36.

Clark, R. C., & Lyons, C. (1999). Using web-based training wisely. *Training,* *36*(7), 51-56.

Clark, R. C., & Mayer, R. E. (2003). *E-Learning and the science of instruction.* San Francisco, CA: Pfeiffer.

Clyde, L. A. (2004). *Weblogs and libraries.* Oxford: Chandos Publishing.

Cohen, V. B. (1985). A reexamination of feedback in computer-based instruction: applications for instructional design. *Educational Technology*, *25*(1), 33-37.

Craik, F. I., & Lockhart, R. S. (1972). Levels of processing: a framework for memory research, *Journal of Verbal Learning and Verbal Behavior*, *11*(6), 671-684.

Curtain, C. (1997). Getting off to a good start on intranets. *Training & Development*, *51*(12), 41-46.

Cyboran, V. (1995). Designing feedback for computer-based training. *Performance & Instruction*, *34*(5), 8-23.

Dempsey, J. V., Driscoll, M. P., & Swindell. L. K. (1993a). Interactive instruction and feedback, chapter text-based feedback. *Educational Technology*, 21-54.

Dempsey, J. V., Driscoll, M. P., & Swindell. L. K. (1993b). Text-based feedback. In J. V. Dempsey & G. C. Sales (Eds.), *Interactive instruction and feedback* (pp. 21-53). Englewood Cliffs, NJ: Educational Technology Publications.

Dick, W., Carey, L., & Carey, J. O. (2005). *The systematic design of instruction* (6th ed). New York: Allyn and Bacon.

Driscoll, M. P. (1998). *Web-based training: Using technology to design adult learning experiences.* San Francisco, CA: Jossey-Bass/ Pfeiffer.

Driscoll, M. P. (2000). *Psychology of Learning for Instruction* (2nd ed). Boston: Allyn and Bacon,

Duffy, T. M., & Jonassen, D. H. (1992). *Constructivism and the technology of instruction.* Hillsdale, NJ: Lawrence Erlbaum Associates, Inc.

Engle, L. (1999). Web-based training as an instructional method: An introduction. Unpublished master's paper, Penn State University, Great Valley School of Graduate Professional Studies, Malvern, Pennsylvania.

Evia, C. (2004). Quality over quantity: a two-step model for reinforcing user feedback in transnational Web-based systems through participatory design. *IEEE Transactions on Professional Communication, 47*(1), 71-74.

Fitch, J. L. (2004). Student feedback in college classroom: A technology solution. *Educational Technology Research & Development, 52* (1), 71-81.

Furnas, G. (1997). Effective view navigation. Retrieved April 27, 2000, from http://www.acm.org/sigchilchi97/proceedings/paper/gwf.htm.

Gagné, R. (1974). Educational technology and the learning process. *Educational Researcher*, *3*(1), 3-8.

Gagné, R. (1970). *The conditions of learning* (2nd ed). New York: Holt, Rinehart and Winston, Inc.

Gagné, R. (1977). *The conditions of learning* (3rd ed). New York: Holt, Rinehart and Winston, Inc.

Gagne, R., & White, R. (1978). Memory structures and learning Outcomes. *Review of Educational Research*, *48*(2),187-222.

Gagne, R. M., Briggs, L. J., & Wager, W. (1987). *Principles of instructional design* (2nd ed). New York: Holt, Rinehart and Winston.

Gagné, R. M., & Medsker, K. L. (1996). *The conditions of learning: Training applications*. New York: Harcourt Brace College Publishers.

Gaynor, P. (1981). The effect of feedback delay on retention of computer-based mathematical material. *Journal of Computer-Based Instruction*, *8*(2), 28-34.

Gilman, D. A. (1969). Comparison of several feedback methods for correcting errors by computer-assisted instruction. *Journal of Educational Psychology*, *60*(6), 503-508.

Gottschalk, P. (1999). Implementation predictors of strategic information systems plans. *Information and Management*, *36*(2), 77-91.

Hall, B. (1997). *Web-based training cookbook*. New York: Wiley & Sons.

Harasim, L., Hiltz, S. R., Teles, L., & Turoff, M. (1995). *Learning networks: A field guide to teaching and learning online*. Cambridge, MA: The MIT Press.

Henke, H. (1997). Evaluating web-based instructional design. Retrieved November 14, 1999, from http://scis.nova.edu/~henkeh/story1.htm.

Hodes, C. L. (1984). Relative effectiveness of corrective and noncorrective feedback in computer assisted instruction on learning and achievement. *Journal of Educational Technology Systems, 13* (4), 249-254.

Jonassen, D. H. (1996). *Computers in the classroom: Mindtools for critical thinking.* Englewood Cliffs, NJ: Prentice-Hall.

Jonassen, D. H. (1996). *Handbook of Research for Educational Communications and Technology.* New York: Simon & Schuster Macmillian.

Jones, M. G., & Okey, J. R. (1995). Interface design for computer-based learning environments. Retrieved November 14, 1999, from http://hbg.psu.edu/bsedlintroldocs/idguide.

Khan, B. H. (1997). *Web-based instruction (Ed.).* Englewood Cliffs, NJ: Educational Technology Publications.

Kilby, T. (1997). What is web-based training? Retrieved November 1, 1999, from http://www.filename.com/wbt/pages/htm.

Kruse, K. & Keil, J. (2000). *Technology based training.* San Francisco, CA: Jossey-Bass Pfeiffer.

Kulhavy, R. W., & Anderson, R. C. (1972). Feedback procedures and programmed instruction. *Journal of Educational Psychology, 63,* 505-512.

Kulhavy, R. (1977). Feedback in written instruction. *Review of Educational Research, 47*(1), 211-232.

Kulhavy, R. W., & Wager, W. (1993). Feedback in programmed instruction: Historical context and implications for practice. In J. Dempsey & G. Ales (Eds.), *Interactive instruction and feedback* (pp. 3-20). Englewood Cliffs, NJ: Educational Technology Publications.

Kulhavy, R. W., Yekovich, F. R., & Dyer, J. W. (1979). Feedback and content review in programmed instruction. *Contemporary Educational Psychology, 4,* 91-98.

Kulik, J. A., & Kulik, C. L. (1988). Timing of feedback and verbal learning. *Review of Educational Research, 58*(1),79-97.

Lee, C-P., Shen, C-W., & Lee, D. (2008). The effect of multimedia instruction for Chinese learning. *Learning, Media and Technology, 33*(2), 127-138.

Lee, D. (1989). The effects of feedback and second try in computer-assisted instruction for a rule-learning task. Unpublished doctoral dissertation, University of Texas at Austin, Austin, Texas.

Lee, D. (1993). Learning English as a foreign language: The case for computer simulation. *British Journal of Educational Technology, 24*(3), 221-222.

Lee, D. (1994). Effects of feedback types and questioning strategies on error correction in computer-based instruction. *Journal of Research in Education, 4*(1), 37-42.

Lee, D. (2004). Web-based instruction in China: Cultural and pedagogical implications and challenges. *Educational Technology Research & Development, 52*(1), 82-85.

Lee, D., & Borland, S. (2007). Implementing computer-supported learning in corporations: challenges and solutions. In F. M. Mdes Neto (Ed.), *Advances in Computer-Supported Learning* (pp. 228-249). Hershey, PA: Idea Group Inc.

Lee, D., Chalmers, T., & Ely, T. (2005). Web-based training in corporations: Design issues. *International Journal of Instructional Media, 31*(4), 345-354.

Lee, D., & Dwyer, F. M. (1994). The effects of varied feedback strategies on students' cognitive and attitude development. *International Journal of Instructional Media, 21*(1), 13-21.

Lee, D., & Frenzelas, G., & Anders, C. (in press). Blended learning for employee training: Influencing factors and important considerations. *International Journal of Instructional Media.*

Lee, D., & Riordan, S. (2007). Handheld Computers in K-12 Schools: Uses, benefits and concerns, *Journal of Leader University, 4* (2), 114-124.

Lee, D., Savenye, W., & Smith, P., (1991). The effects of feedback and second try on error correction. Proceedings of Selected Research and Development Presentations at the 1991 National Convention of the Association for Educational Communications and Technology, 440-461.

Lockard, J., Abrams, P. D., & Many, W. A. (1990). *Microcomputers for educators* (2nd ed). Glenview, IL: Scott Foresman/Little Brown Higher Education.

Lublin, S. C. (1965). Reinforcement schedules, scholastic aptitude, autonomy need and achievement in a programmed course. *Journal of Educational Psychology, 56*, 295-302.

Macpherson, A., Homan, G., & Wilkinson, K. (2005). The implementation and use of e-learning in the corporate university. *Journal of Workplace Learning, 17*(1/2), 33-48.

Mason, B. J., & Bruning, R. (2001). Providing feedback in computer-based instruction: What the research tells us. Retrieved 2004, from http://dwb.unl.edu/Edit/MB/MasonBruning.html.

Mayer, R. E. (1975). Different problem-solving competencies established in learning computer programming with or without meaningful models. *Journal of Educational Psychology, 67*(6), 725-734.

Mayer, R. E. (1976). Some conditions of meaningful learning for computer programming advanced organizer and subject control of frame order. *Journal of Educational Psychology, 68*(2), 143-150.

McCreary, E. K., & Van Duran, J. (1987). Educational application of computer conferencing. *Canadian Journal of Educational Communication, 16*(2), 107-115.

Moore, J. W., & Smith, W. (1964). Role of knowledge of results in programmed instruction. *Psychological Report, 14*, 407-423.

More, A. J. (1969). Delay of feedback and the acquisition and retention on verbal materials in the classroom. *Journal of Educational Psychology, 60*, 339-342.

Mory, E. H. (1991). The effects of adaptive feedback on student performance, feedback study time, and lesson efficiency within computer-based instruction. Unpublished doctoral dissertation, Florida State University, Tallahassee, Florida.

Mory, E. H. (1996), Feedback research. In D. H. Jonassen (Ed.), *Handbook of Research for Educational Communications and Technology*, (pp. 919-956). New York: Simon and Schuster Macmillan.

Mory, E. H. (1994). Adaptive feedback in computer-based instruction: Effects of response certitude on performance, feedback-study time, and efficiency. *Journal of Educational Computing Research, 11*(3), 263-290.

Nisar, T. (2004). E-learning in public organizations. *Public Management, 33*(1), 79-88.

Palloff, R. M., & Pratt, K. (1999). *Building learning communities in cyberspace: Effective strategies for the online classroom.* San Francisco, CA: Jossey-Bass Publishers.

Perry, D. (2003). Handheld computers (PDAs) in schools. Retrieved January 26, 2006, from http://publications.becta.org.uk/display.efm?resID=25833,2003.

Phye, G. D. (1979). The processing of informative feedback about multiple choice test performance. *Contemporary Educational Psychology, 4*, 381- 394.

Polsson, K. (2005). Chronology of handheld computers, Retrieved January 25, 2006, from http://www.islandnet.com/~kpolsson/handheld.

Polya, G. (1957). *How to solve it* (2nd ed). New York: Doubleday.

Polyson, S., Saltzberg, S., & Goodwin-Jones, R., (1996). A practical guide to teaching with the World Wide Web. *Syllabus, 10*(2), 12-16.

Pridemore, D. R., & Klein, J. R. (1991). Control of feedback in computer-assisted instruction. *Educational Technology Research & Development, 39*(4), 5-128.

Pridemore, D. R., & Klein, J. R. (1995). Control of practice and level of feedback in computer-based instruction. *Contemporary Educational Psychology, 20*(4), 444-450.

Rankin, R. J., & Trepper, T. (1978). Retention and delay of feedback in computer-assisted instruction task. *Educational Technology Research & Development, 48*(3), 5-22.

Rieber, L. P. (1992). Computer-based microworlds: A bridge between constructivism and direct instruction. *Educational Technology Research & Development, 41*(1), 93-106

Riordan, S., & Lee, D. (2007). Handheld computers in K-12 schools: Uses, benefits and concerns. *Journal of Leader University, 4*(2), 114-124.

Roberts, F. C., & Park, O. (1984). Feedback strategies and cognitive style in computer-based instruction. *Journal of Instructional Psychology, 11*, 63-74.

Roper, W. J. (1977). Feedback in computer-assisted instruction. *Programmed Learning and Educational Technology, 14*, 43-49.

Roussou, M., Oliver, M., & Slater, M. (2006). The virtual playground: an educational virtual reality environment for evaluating interactivity and conceptual learning. *Journal of Virtual Reality, 10* (3-4), 227-240.

Sassenrath, J. M. (1972). Alpha factor analyses of reading measures at the elementary, secondary, and college levels. *Journal of Reading Behavior, 5*(4),1972-1973.

Sassenrath, J. M., (1975). Theory and results on feedback and retention. *Journal of Educational Psychology, 67*, 894-899.

Sassenrath, J. M., & Yonge, G. D. (1968). Delayed information feedback, feedback cues, retention set, and delayed retention. *Journal of Educational Psychology, 59*(2), 69-73.

Sassenrath, J. M., & Yonge, G. D. (1969). Effects of delayed information feedback and feedback cues in learning and retention. *Journal of Educational Psychology, 60*(3), 174-177.

Schimmel, B. (1988). Providing meaningful feedback in courseware. In D. Jonassen (Ed.), *Instructional designs for microcomputer courseware* (pp.183-195). Hillsdale, NJ: Lawrence Erlbaum Associates.

Schmidt, R. A., Young, D. E., Swinnen, S., & Shapiro, D. C. (1989). Summary knowledge of results for skill acquisition: Support for the guidance hypothesis. *Journal of Experimental Psychology: Learning, Memory, and Cognition, 15*(2), 352-359.

Siegel, M. A., & Misselt, A. (1984). Adaptive feedback and review paradigm for computer-based drills. *Journal of Educational Psychology, 76*(2), 310-317.

Shelly, G. B., Cashman, T. J., Gunter, R. E., & Gunter, G. A. (2004). *Teachers discovering computers-integrating technology into the classroom*. Boston, MA: Thomson Course Technology.

Skinner, B. F. (1951) How to teach animals. *Scientific American, 185,* 26-29.

Skinner, B. F. (1957). *Verbal behavior.* New York: Appleton-Century Crofts.

Skinner, B. F. (1959). The programming of verbal knowledge. In E. Galanter (Ed.), *Automatic teaching: The state of the art* (pp. 63-68). New York: John Wiley & Sons.

Skinner, B. F. (1968). *The technology of teaching*. New York: Appleton Century-Crofts.

Smith, K. U., & Smith, M. F. (1966). *Cybernetic principles of learning and educational design*. New York: Holt, Rinehart and Winston.

Smith, P. L., & Ragan, T. J. (1999). *Instructional Design*. John Wiley & Sons, Inc.

Smith, S., Tyler, M., & Benscoter, A. (1999). Internet supported teaching: advice from the trenches. Retrieved June 6, 2000, from http://www.usdla.org/Ed_magazine/illuminactive/Jan_1999/internet.htm.

Spiro, R. J., Feltovich, P. J., Jacobson, M. J., & Coulson, R. L. (1991). Cognitive flexibility, constructivism, and hypertext: random access instruction for advanced knowledge acquisition in ill-structured domains. *Educational Technology*, 14-33.

Spock, P. (1988). The effect of feedback, correctness of response, and confidence of response on a rule-learning task using computer-assisted instruction, Unpublished Doctoral Dissertation, the University of Texas at Austin, Austin, TX.

Sternberg, R. J. (1983). Criteria for intellectual skills training. *Educational Researcher*, 12(2), 6-12.

Sturges, P. T. (1969). Verbal retention as a function of the in formativeness and delay of information feedback. *Journal of Educational Psychology*, *60*, 11-14.

Surber, J. R., & Anderson, R. C. (1975). Delayed-retention effect in natural classroom settings. *Journal of Computer-Based Instruction*, *12*(1), 17-20.

Thorndike, E. L. (1911). *Animal intelligence.* New York: Macmillan.

Trombley, B. K., & Lee, D. (2002). Web-based learning in corporations: who is using it and why, who is not and why not? *Journal of Educational Media*, *27*(3), 137-146.

Vaughan, T. (1996). *Multimedia: Making it work*, (3rd ed). Berkeley, CA: Osborne McGraw-Hill.

Waddick, J. (1994). Case study: The creation of a computer learning environment as an alternative to traditional lecturing methods in chemistry. *Educational and Training Technology International*, *31*(2), 98-103.

Wager, W., & Mory, E. H. (1993). The role of questions in learning. In J. V. Dempsey, & G. C. Sales (Eds.), *Interactive Instruction and Feedback* (pp. 55-74). Englewood Cliffs, NJ: Educational Technology Publications.

Ward, S. (1998). Secret recipe for CBT. *Technical Training*, *9*(5), 16-22.

Webster's New World Dictionary. (1976). New York: Random House.

Williams, R. (1998). Challenges to the optimal delivery of a training program via the World Wide Web. Retrieved September 23, 1999 from http://www.trainingplace.com/source/wbtlimit.html.

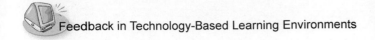

Wlodowski, R. J. (1985). *Enhancing adult motivation to learn*. San Francisco, CA: Jossey-Bass Publishers.

Yoon, S., Ho, F., & Hedburg, J. (2005). Teacher understandings of technology affordances and their impact on the design of engaging learning experiences. *Educational Media International, 42*(4), 297-316.

Index
索引

H

Hall, B., 149

handheld computers, 15, 154

Harasim, L., 149

hard disk, 13

hard drive, 12, 13

hardware, 6, 10, 11, 15, 18, 39, 41, 134, 135

hatrooms, 22, 27

headset, 15

Hedburg, J., 159

Henke, H., 149

Hiltz, S. R., 149

Ho, F., 159

Hodes, C. L., 149

Homan, G., 152

human control mechanism, 52

human-machine-human learning, 32

hyper-link, 38

hypermedia, 18

hypertext, 18, 60, 158

hypothesis, 92, 113, 157

I

image, 14

image scanner, 14

in-house, 30

input devices, 11, 14, 140

instructional designers, 118, 120

instructional games, 19

instructor, 3, 17, 19, 24, 28, 32, 36, 37, 38, 39, 128, 135

instructor-led, 3, 28

interaction effects, 80, 91

interface, 12, 38, 42, 126, 134

internet, 2, 6, 8, 10, 14, 21, 22, 27, 31, 101, 133, 134, 158

interpersonal relations, 32

intranet, 2, 3

J

Jacobson, M. J., 158

Jonassen, D. H., 147, 149

Jones, M. G., 149

Joystick, 14, 98

K

KCR, 46, 48, 71, 72, 73, 78, 80, 81, 82, 83, 86, 87, 91, 110, 112, 113, 114, 120, 128

KCR feedback, 48, 71, 73, 78, 80, 81, 83, 86, 87, 91, 110, 113, 120, 128

Keil, J., 150

keyboard, 11, 14, 133

Khan, B. H., 150

Kilby, T., 150

Klein, J. R., 155

knowledge of response, 46, 76

knowledge of the correct answer, 69, 71, 72, 74

KOR, 46, 48, 73, 75, 82, 83, 86, 87, 91, 93, 109, 110, 111, 112, 113, 120, 128, 139

Kruse, K., 150

Kulhavy, R. W., 143, 150

Kulik, C. L., 150

應用科學類　AB0008

Feedback in Technology-Based Learing Environments
回饋訊息於科技教學的功能與效應

作　　者 / 李宜珍（Doris Lee, Ph.D.）
插　　圖 / 林浩寬（Andrew Lin）
發 行 人 / 宋政坤
執行編輯 / 林世玲
圖文排版 / 鄭維心
封面設計 / 李孟瑾
數位轉譯 / 徐真玉　沈裕閎
圖書銷售 / 林怡君
法律顧問 / 毛國樑　律師
出版印製 / 秀威資訊科技股份有限公司
　　　　　台北市內湖區瑞光路 583 巷 25 號 1 樓
　　　　　電話：02-2657-9211　　　傳真：02-2657-9106
　　　　　E-mail：service@showwe.com.tw
經 銷 商 / 紅螞蟻圖書有限公司
　　　　　台北市內湖區舊宗路二段 121 巷 28、32 號 4 樓
　　　　　電話：02-2795-3656　　　傳真：02-2795-4100
　　　　　http://www.e-redant.com

2008 年 10 月 BOD 一版
定價：180 元

讀　者　回　函　卡

感謝您購買本書，為提升服務品質，煩請填寫以下問卷，收到您的寶貴意見後，我們會仔細收藏記錄並回贈紀念品，謝謝！

1.您購買的書名：_____

2.您從何得知本書的消息？

　　□網路書店　　□部落格　　□資料庫搜尋　　□書訊　　□電子報　　□書店

　　□平面媒體　　□ 朋友推薦　　□網站推薦　　□其他_____

3.您對本書的評價：(請填代號　1.非常滿意 2.滿意 3.尚可 4.再改進)

　　封面設計____　版面編排____　內容____　文/譯筆____　價格____

4.讀完書後您覺得：

　　□很有收獲　　□有收獲　　□收獲不多　　□沒收獲

5.您會推薦本書給朋友嗎？

　　□會　□不會，為什麼？_____

6.其他寶貴的意見：_____

_____

_____

_____

讀者基本資料

姓名：_____　年齡：_____　性別：□女　□男

聯絡電話：_____　E-mail：_____

地址：_____

學歷：□高中(含)以下　　□高中　　□專科學校　　□大學

　　　□研究所(含)以上 □其他_____

職業：□製造業 □金融業 □資訊業 □軍警 □傳播業 □自由業

　　　□服務業 □公務員 □教職　　□學生 □其他_____

To：114

台北市內湖區瑞光路 583 巷 25 號 1 樓

秀威資訊科技股份有限公司　　　收

寄件人姓名：

寄件人地址：□□□

(請沿線對摺寄回,謝謝!)

秀威與 BOD

BOD（Books On Demand）是數位出版的大趨勢，秀威資訊率先運用 POD 數位印刷設備來生產書籍，並提供作者全程數位出版服務，致使書籍產銷零庫存，知識傳承不絕版，目前已開闢以下書系：

一、BOD 學術著作—專業論述的閱讀延伸
二、BOD 個人著作—分享生命的心路歷程
三、BOD 旅遊著作—個人深度旅遊文學創作
四、BOD 大陸學者—大陸專業學者學術出版
五、POD 獨家經銷—數位產製的代發行書籍

BOD 秀威網路書店：www.showwe.com.tw
政府出版品網路書店：www.govbooks.com.tw

永不絕版的故事‧自己寫‧永不休止的音符‧自己唱